Problems and
Solutions for
General
Chemistry
and College
Chemistry

Problems and Solutions for
General Chemistry and College Chemistry

FIFTH EDITIONS
BY NEBERGALL,
SCHMIDT,
AND
HOLTZCLAW

John H. Meiser
Ball State University

Frederick K. Ault
Ball State University

Henry F. Holtzclaw, Jr.
University of Nebraska, Lincoln

D. C. Heath and Company
Lexington, Massachusetts Toronto London

Preface

Beginning students of chemistry often need direction beyond that presented in their textbooks, particularly when working with mathematical type exercises. This problems and solutions manual is designed to meet that student need. We believe the manual will increase the problem-solving ability, as well as the conceptual understanding, of those who use it diligently.

Problems and Solutions for General Chemistry and College Chemistry is keyed to the textbooks *General Chemistry* and *College Chemistry, 5th Editions*, by Nebergall, Schmidt, and Holtzclaw. These textbooks clearly and succinctly present the concepts basic to general chemical knowledge and to any further study of chemistry. In addition, there are specific chapters that emphasize the descriptive and qualitative aspects of the elements, particularly their occurrence, abundance, physical properties, common chemical reactions, and economic impact. Other chapters capably introduce organic chemistry, biochemistry, thermodynamics, nuclear chemistry, and electrochemistry to give a comprehensive picture of chemistry for the student.

We have organized this manual as an aid to learning, specifically correlated with the chapters in the Nebergall, et al., textbooks that involve important chemical concepts and also mathematics. Included at the ends of these chapters are numerous problems that illustrate particularly important aspects of these concepts. In this manual there is a short introduction to each chapter and a section on definitions and equations, followed by a detailed solution to at least one example of each type of problem in the chapter. The problems selected for the manual are indicated in the textbooks by the symbol ⑤. The selected problems comprise one-third to one-half of the number of problems in the chapter. The manual presentation includes statements of the problems selected, making it suitable for use as a supplement in any chemistry course that uses the problem approach to teaching.

Problems and Solutions for General Chemistry and College Chemistry also contains a self-instructional section (Part II) that teaches the use of exponential numbers and logarithms. Self-evaluation exercises enable student to check on their progress at appropriate intervals.

We know that many thousands of students have successfully used the textbooks *General Chemistry* and *College Chemistry* through the past four editions in pursuing goals in chemistry and the allied-health sciences. Along with the present edition, use of *Problems and Solutions for General Chemistry and College Chemistry, 5th Editions*, should allow students to more easily reach these goals through a better understanding of the quantitative aspects of the sciences.

In producing this manual, we are indebted to our colleagues for many helpful suggestions, to our typists for the physical body of the manuscript, and to our families particularly for their understanding during the long hours spent in preparing this work.

J. H. Meiser

F. K. Ault

H. F. Holtzclaw, Jr.

Contents

PART ONE
Problems and Solutions

UNIT I

Some Fundamental Concepts

1

INTRODUCTION

This chapter of the textbooks includes a brief introduction to the nature of matter and energy and to the units of measurement essential for quantitatively expressing observations of scientific phenomena. The metric system of measurement is emphasized throughout.

The problem-solving exercises stress development of skills for treating each of the following concepts: expression of numbers in scientific notation; conversion among metric prefixes; expressions of mass-to-volume ratios, as density and specific gravity; conversion of measurements expressed in English units to appropriate metric units; and conversion of temperature measurement from one scale to another. Calculations involving heat are also included.

DEFINITIONS AND FORMULAS

Density (D) The density of a substance is the ratio of its mass, M, to its volume, V; for a solid or liquid, density is expressed as mass (in grams) per cubic centimeter or milliliter:

$$\text{Density} = \frac{\text{mass}}{\text{volume}} = \frac{\text{grams}}{\text{ml or cm}^3}$$

Specific gravity (sp gr) The specific gravity of a substance, as understood in chemistry, is the ratio of the density of an unknown substance to the density of a substance used as reference standard (both measured at the same temperature). For solids and liquids, water is most often the standard; for gases, air is most frequently the standard.

$$\text{Specific gravity (solid or liquid)} = \frac{\text{density (solid or liquid)}}{\text{density water}}$$

Since the density of water is approximately 1.0 g/ml over a range of temperatures near room temperature, the specific gravity of a substance has approximately the same numerical value as the density of the substance; specific gravity is unitless because the units of density divide out:

$$\text{sp gr} = \frac{D \text{ (unknown)}}{D \text{ (standard)}} = \frac{5.0 \text{ g/ml}}{1.0 \text{ g/ml}} = 5.0$$

Heat   Heat is a form of energy. It usually is expressed in small calories. Heat gain or heat loss is calculated by the following relation:

$$\text{Heat} = \text{mass substance} \times \text{specific heat} \times \text{temp. change } (\Delta t)$$

The transfer of heat from one body to another is expressed through an application of the above equation. Heat lost = heat gained. Therefore,

$$(m_1)\,(\text{sp heat}_1)\,(\Delta t_1) = (m_2)\,(\text{sp heat}_2)\,(\Delta t_2)$$

Metric conversion relations

Length:

$$km \underset{1000 \div}{\overset{\times 1000}{\rightleftarrows}} m \underset{10 \div}{\overset{\times 10}{\rightleftarrows}} dm \underset{10 \div}{\overset{\times 10}{\rightleftarrows}} cm \underset{10 \div}{\overset{\times 10}{\rightleftarrows}} mm$$

$$1 \text{ cm} = 10^8 \text{ Å, or } 1 \text{ Å} = 10^{-8} \text{ cm}$$

Mass:

$$kg \underset{1000 \div}{\overset{\times 1000}{\rightleftarrows}} g \underset{1000 \div}{\overset{\times 1000}{\rightleftarrows}} mg \underset{1000 \div}{\overset{\times 1000}{\rightleftarrows}} \mu g$$

Volume:

$$m^3 \underset{1000 \div}{\overset{\times 1000}{\rightleftarrows}} l \underset{1000 \div}{\overset{\times 1000}{\rightleftarrows}} cm^3 \text{ or ml} \underset{1000 \div}{\overset{\times 1000}{\rightleftarrows}} \mu l$$

PROBLEMS

Note: Reference to tables, figures, and appendices found in the problem statements are to tables and figures in the fifth editions of *General Chemistry* and *College Chemistry*, by Nebergall, et al.

1. Express 4.71 kg in grams, milligrams, metric tons, and pounds.

    Soln: No. g = 4.71 kg × 10^3 g/kg = 4.71 × 10^3 g
               No. mg = 4.71 kg × 10^6 mg/kg = 4.71 × 10^6 mg

$$\text{No. m-tons} = 4.71 \text{ kg} \times \frac{1 \text{ m-ton}}{10^3 \text{ kg}} = 4.71 \times 10^{-3} \text{ m-tons}$$

$$\text{No. lb} = 4.71 \text{ kg} \times \frac{1 \text{ lb}}{0.4536 \text{ kg}} = 10.4 \text{ lb}$$

3. Calculate the number of liters in a gallon and in a cubic inch.

Soln: No. liters = 4 qt/gal × 0.9463 liter/qt = 3.785 liters
1 cu in. = $(2.54 \text{ cm})^3$ = 16.4 cm^3

$$\text{No. liters} = 16.39 \text{ cm}^3/\text{in.}^3 \times \frac{1.0 \text{ liter}}{10^3 \text{ cm}^3} = 1.639 \times 10^{-2} \text{ liter}$$

4. A rectangular tank which measures 500 cm in width and 5.00 m in length is filled with water to a depth of 100 mm. What volume of water is required? What mass of water is required?

Soln: Convert all measurements to cm and calculate the volume of the box in cubic centimeters.

Volume = 500 cm × 500 cm × 10.0 cm = 2.50×10^6 cm^3
2.50×10^6 cm^3 × $1/10^3$ cm^3 = 2.50×10^3 liters
Since one liter of water weighs approximately 1 kg, the total mass is 2.50×10^3 liters × 1 kg/liter = 2.50×10^3 kg.

6. How many millimeters are in one yard? How many yards are in one centimeter? How many Angstrom units are in one yard?

Soln: No. mm = 36 in./yd × 2.540 cm/in. × 10 mm/cm = 9.144×10^2 mm

$$\text{No. yd} = \frac{1 \text{ yd}}{36 \text{ in.}} \times \frac{1 \text{ in.}}{2.540 \text{ cm}} = 1.094 \times 10^{-2} \text{ yd}$$

$$\text{No. Å} = 36 \text{ in./yd} \times 2.540 \text{ cm/in.} \times \frac{1 \text{ Å}}{10^{-8}} \text{ cm} = 9.144 \times 10^9 \text{ Å}$$

9. If milk is sold for 85 cents per half-gallon, what is its cost per liter?

Soln:
$$\text{Cost/liter} = \frac{\text{half-gal}}{2 \text{ qt}} \times \frac{1 \text{ qt}}{0.9463 \text{ liter}} \times \frac{85 \text{ cents}}{1 \text{ half-gal}} = 45 \text{ cents}$$

10. What is the volume of 6.75 g of mercury? of 27.0 g of mercury? of 362 g of mercury? (The density of mercury is 13.5 g/cm^3.)

Soln:
$D = \dfrac{M}{V}$ Substituting values into the density expression yields
13.5 g/cm^3 = 6.75 g/V

$$V = \frac{6.75 \text{ g}}{13.5 \text{ g/cm}^3} = 0.500 \text{ cm}^3$$

11. (a) The world record for the 100-meter dash is 9.90 seconds. If a runner could maintain this pace for a mile, what would be his time in minutes for the mile? (b) What would be the time in seconds for running the 100-meter dash if it were run at the average speed required for a 4.00-minute mile?

Soln: (a) 1 mile = 1.609 km = 1.609×10^3 m

$$\text{Time} = 1.609 \times 10^3 \text{ m/mile} \times \frac{9.90 \text{ sec}}{10^2 \text{ m}} \times \frac{1 \text{ min}}{60 \text{ sec}} = 2.65 \text{ min}$$

13. If 5.84×10^{13} helium atoms (spherical) are laid in a line, each touching the next, they measure 6.75 miles. What is the diameter of a helium atom in Å?

Soln: Length ÷ no. of particles = length/particle

$$\text{No. Å} = 1.609 \text{ km/mile} \times 6.75 \text{ miles} \times 10^3 \text{ m/km} \times \frac{1 \text{ Å}}{10^{-10} \text{ m}} \times$$

$$\frac{1}{5.84 \times 10^{13}} = 1.86 \text{ Å}$$

14. A student has two beakers. The sum of their masses is 264.2 grams. The difference in their masses is 11.40 grams. What is the mass of each beaker?

Soln: Let the mass of one beaker be A and of the other, B. Then $A + B = 264.2$ and $A - B = 11.40$. Add the equations:

$$\begin{aligned} A + B &= 264.2 \\ A - B &= \ \ 11.40 \\ \hline 2A \quad\ \ &= 275.6 \end{aligned}$$

Therefore, $A = 137.8$ g.

Substitute the value for A in either equation to find B.

$$137.8 + B = 264.2$$
$$B = 264.2 - 137.8$$
$$= 126.4 \text{ g}$$

16. Calculate the density of a solid for which 40.0 g occupies a volume of 21.7 cm^3.

Soln: $$D = \frac{M}{V} = \frac{40.0 \text{ g}}{21.7 \text{ cm}^3} = 1.84 \text{ g/cm}^3$$

18. What is the specific gravity of a liquid if 500 cm^3 weighs as much as 850 cm^3 of water?

Soln: $$\text{Specific gravity} = \frac{M_1/V_1}{M_{H_2O}/V_{H_2O}}; \quad M_1 = M_{H_2O}$$

$$\text{Specific gravity} = \frac{V_{H_2O}}{V_1} = \frac{850 \text{ cm}^3}{500 \text{ cm}^3} = 1.70$$

19. What is the weight of each of the following?
    (a) 33.3 cm^3 of a liquid, density = 1.836 g/cm^3.
    (b) 50.0 cm^3 of a liquid, specific gravity = 0.790.
    (c) 4.00 cm^3 of mercury, density = 13.5 g/cm^3.

Soln: $D = M/V$, or $M = DV$
    (a) $M = 1.836 \text{ g/cm}^3 \times 33.3 \text{ cm}^3 = 61.1 \text{ g}$

20. Use the following data to calculate the specific gravity of an unknown liquid:

> Weight of pycnometer + liquid: 16.7539 g
> Weight of pycnometer empty: 10.6402 g
> Weight of pycnometer + water: 14.7572 g

Soln:

$$\text{Wt of liquid} = 6.1137 \text{ g}$$
$$\text{Wt of same volume of water} = 4.1170 \text{ g}$$
$$\text{Specific gravity} = \frac{M_1/V}{M_{H_2O}/V} = \frac{6.1137 \text{ g}}{4.1170 \text{ g}} = 1.4850$$

23. A piece of lead when put into water displaces 119.0 cubic centimeters. The specific gravity of pure lead is 11.3. Calculate the weight of the piece of lead.

Soln: The density of water is one gram per cubic centimeter. Then 119.0 cm^3 of water weighs 119.0 g; and since lead is 11.3 times as heavy as water, the weight for an equal volume of lead is

$$119.0 \text{ g} \times 11.3 = 1.34 \times 10^3 \text{ g}$$
$$= 1.34 \text{ kg}$$

24. The density of uranium is 18.7 g/cm^3. Which of the following contains the greatest mass of uranium: 1.0 lb; 0.50 kg; 0.050 liter?

Soln: First convert all samples to common units of grams.
Sample 1: 1 lb = 453.6 g
Sample 2: 0.50 kg $\times$ 10^3 g/kg = 500 g
Sample 3: 0.050 liter $\times$ 10^3 cm^3/liter $\times$ 18.7 g/cm^3 = 935 g
Therefore, Sample 3 contains the greatest mass.

26. The density of 96.0% sulfuric acid is 1.84 g/ml. How many milliliters of volume will be occupied by 25.0 g of the acid? How many grams of water will be present in this volume?

Soln: Since $D = M/V$, $V = M/D$.

$$V \text{ acid} = 25.0 \text{ g} \times \frac{1 \text{ ml}}{1.84 \text{ g}} = 13.6 \text{ ml}$$

The sulfuric acid is 96.0%. Therefore, water is 4.0% of 25.0 g.

$$25.0 \text{ g} \times 0.04 = 1.00 \text{ g}.$$

28. Make the following temperature conversions:
(a) 0° F to degrees Celsius.   (d) −40° C to degrees Fahrenheit.
(b) 37.0° C to degrees Fahrenheit. (e) 150.0° C to degrees Fahrenheit.
(c) 66.0° F to degrees Kelvin.   (f) 7.00° K to degrees Fahrenheit.

Soln: (b) $F = \frac{9}{5}C + 32 = \frac{9}{5}(37.0) + 32 = 66.6 + 32 = 98.6°F$
(c) First convert °F to °C.

$$C = \tfrac{5}{9}(F - 32) = \tfrac{5}{9}(66.0 - 32.0) = 18.9° \text{ C}$$
$$K = °C + 273.1 = 18.9 + 273.1 = 292° \text{ K}$$

30. How many calories of heat would be required to increase the temperature of 600 g of water from 80.6° F to 107.6°F?

Soln: First convert the temperatures to Celsius.

$$C = \tfrac{5}{9}(F - 32)$$

°F to °C: 107.6° F = 42.0° C
         80.6° F = 27.0° C
Temperature change = 42.0 − 27.0 = 15.0° C
Heat required = mass × specific heat × temperature change
         = 600 g × 1 cal/g° C × 15.0° C = 9000 cal

32. If 360 calories are added to 30.0 g of water at 298° K, what is the resulting temperature?

Soln: Heat added = 360 cal; therefore the relation for temperature change is

Heat = mass × specific heat × temperature change
360 cal = 30 g × 1 cal/g° C × ΔT

The units divide out as shown:

$$\Delta T = \frac{360}{30} = 12°\text{C}$$

$$T_f = 298° + 12° = 310°$$

33. Calculate the heat capacity of the following:
(a) 22.6 g of water (specific heat, 1.00 cal/g °C).
(b) 1.10 kg of aluminum (specific heat, 0.01 cal/g °C).
(c) 50.0 ml of copper (density, 8.94 g/cm³, specific heat, 0.09 cal/g °C).

Soln: Heat capacity = grams × specific heat
(a) Heat capacity = 22.6 g × 1.00 cal/g °C
                  = 22.6 cal/°C
(c) First convert volume of copper to a mass:

Mass = 50.0 ml × 8.94 g/cm³ × cm³/ml = 447 g
Heat capacity = 447 g × 0.09 cal/g °C = 40 cal/°C

35. How many grams of water at 25° C must be combined with 1.50 liters of water at 100° C so that the resulting combination will have a temperature of 70° C?

Soln: Grams$_1$ × specific heat$_1$ × $|\Delta t_1|$ = grams$_2$ × specific heat$_2$ × $|\Delta t_2|$
grams at 25° × 1 cal/g °C × |(70 − 25)| = 1500 g ×
1 cal/g °C × |(70 − 100)|

6

$$45 \text{ g} = 1500 \times 30$$
$$\text{water at } 25° \text{ C} = 1000 \text{ g}$$

37. When 18 g of water is formed from its elements, 68,315 cal of heat is evolved. What would be the change in temperature of a piece of zinc weighing 10.0 kilograms if heated by the quantity of heat liberated by the formation of 18.0 g of water? (The specific heat of zinc is 0.093 cal per gram per degree.)

Soln: Heat = grams × specific heat × temperature change

$$68,315 \text{ cal} = 1.0 \times 10^4 \text{ g} \times 0.093 \text{ cal/g }°\text{C} \times \Delta t$$

$$\Delta t = \frac{68.3 \times 10^3}{1.0 \times 10^4 \times 0.093} = 73° \text{ C}$$

40. A spherical ball of iron weighs one metric ton. What would be its diameter? (The specific gravity of iron is 7.86.)

Soln: Volume of a sphere is $\frac{4}{3}\pi r^3$.

1 metric ton = 1000 kg = 1.00×10^6 g

$$\text{Volume} = 1.00 \times 10^6 \text{ g} \times \frac{1 \text{ cm}^3}{7.86 \text{ g}} = 1.27 \times 10^5 \text{ cm}^3$$

$$V = 1.27 \times 10^5 \text{ cm}^3 = \frac{4}{3}\pi r^3$$

$$r^3 = 3 \times 1.27 \times \frac{10^5}{4\pi} = 3.04 \times 10^4 \text{ cm}^3$$

$$r = 31.2 \text{ cm}, \quad d = 2r = 62.4 \text{ cm}$$

UNIT II

Symbols, Formulas, and Equations

2

INTRODUCTION

The interrelations of symbols, formulas, and equations are treated in this chapter. A chemical formula defines a substance by its constituent parts — the kind and the ratio of atoms that make up a substance. A chemical equation is a shorthand description of a chemical reaction that provides both qualitative and quantitative information about the substances involved.

Calculations involving formulas and equations include deriving formulas for substances, determining weights of substances involved in reactions, and using the mole to compare quantities of matter by the number of particles present. An operational definition of the mole is given below for ready reference as needed for the problems.

FORMULAS AND DEFINITIONS

Formula weight (FW)   Sum of the atomic weights of all the atoms shown by the formula of a substance.

Molecular weight (MW)   Formula weight of a substance that exists as discrete molecules.

Gram-atomic weight (GAW)   Atomic weight of an element expressed in grams.

Gram-formula weight (GFW)   Formula weight of a substance expressed in grams.

Gram-molecular weight (GMW)   Molecular weight of a substance expressed in grams.

Mole Mass of a substance, characteristic for that substance, that contains Avogadro's number (6.022×10^{23}) of particles. Therefore, a mole is equivalent to:

Number of atoms in a gram-atomic weight (GAW) of an elemental substance, for example, in 23.0 g Na.

Number of formula units in a gram-formula weight (GFW) of an ionic substance, for example, in 58.5 g NaCl.

Number of molecules in a gram-molecular weight (GMW) of a molecular substance, for example, in 44.0 g CO_2.

$$\text{No. moles of a substance} = \frac{\text{weight of the substance}}{\text{GFW, GMW, or GAW of the substance}}$$

PROBLEMS

1. Calculate the formula weight of each of the following compounds:
   (a) $C_2 H_4 Cl_2$.

Soln:

Atomic weight	X	no. atoms		
C = 12.01	X	2	=	24.02
H = 1.01	X	4	=	4.04
Cl = 35.45	X	2	=	70.90
FW of $C_2 H_4 Cl_2$			=	98.96

   (d) $[Co(NH_3)_6]Br_3$.

Soln:

Atomic weight	X	no. atoms		
Co = 58.933	X	1	=	58.933
N = 14.0067	X	6	=	84.040
H = 1.00797	X	18	=	18.143
Br = 79.909	X	3	=	239.727
FW of $[Co(NH_3)_6]Br_3$			=	400.8

3. Calculate the formula weight of each of the following minerals:
   (a) Carnotite, $K_2(UO_2)_2(VO_4)_2 \cdot 3H_2O$.

Soln:

Atomic weight	X	no. atoms		
K = 39.102	X	2	=.	78.20
U =238.04	X	2	=	476.08
O = 15.9994	X	15	=	239.99
V = 50.942	X	2	=	101.88
H = 1.00797	X	6	=	6.05
FW of $K_2(UO_2)_2(VO_4)_2 \cdot 3H_2O$			=	902.2

4. What is the weight of the following in grams?
   (b) Four gram-formula weights of magnesium oxide.

9

Soln: GFW of magnesium oxide, MgO
= GAW Mg + GAW O
= 24.305 g + 15.9994 g = 40.304 g
Wt MgO = 4 GFW × 40.304 g/GFW = 161.22 g.

(d) One millimole of sulfur dioxide.

Soln: One mole SO_2 = GAW (S + 2(O))
= 32.1 g + 2(16.0 g) = 64.1 g
Wt SO_2 = 64.1 g/mole × 1 mole/1000 m-moles = 6.41 × 10^{-2} g.

(e) 5.25 × 10^{23} molecules of hydrogen chloride.

Soln: GFW of HCl = GAW of H + GAW of Cl
= 1.01 g + 35.5 g = 36.5 g

$$Wt\ HCl = (5.25 \times 10^{23}\ molecules) \times \frac{1\ mole}{6.022 \times 10^{23}\ molecules}$$

$$\times\ 36.5\ g/mole = 31.8\ g$$

5. Calculate the number of moles in:
   (b) 315 g $Mg(HCO_3)_2$.

Soln: GFW of $Mg(HCO_3)_2$ = GAW of (Mg + 2H + 2C + 6O)
= 24.3 g + 2.0 g + 24.0 g + 96.0 g = 146.3 g

$$No.\ moles = 315\ g \times \frac{1\ mole}{146.3\ g} = 2.15\ moles$$

(d) 3.6 mg of NH_3.

Soln: GFW of NH_3 = GAW of (N + 3H) = 14.0 g + 3.0 g = 17.0 g

$$No.\ moles = 3.6\ mg \times \frac{1\ g}{1000\ mg} \times \frac{1\ mole}{17.0\ g} = 0.00021\ mole$$

8. Which of the following amounts contains the greatest number of atoms of any type: 2.0 grams of lithium metal, 0.30 gram-atom of hydrogen, one-sixth mole of hydrogen molecules, 2.0 × 10^{23} atoms of mercury, or 150 millimoles of bromine molecules? Show why.

Soln:
$$No.\ atoms\ Li = 2.0\ g \times \frac{1\ GAW}{6.9\ g} \times \frac{6.02 \times 10^{23}\ atoms}{GAW} = 1.74 \times 10^{23}.$$

$$No.\ atoms\ H = 0.30\ g\text{-}atom \times \frac{1\ GAW}{g\text{-}atom} \times \frac{6.02 \times 10^{23}\ atoms}{GAW}$$
$$= 1.8 \times 10^{23}.$$

$$No.\ atoms\ H = \frac{1}{6}\ mole\ H_2 \times \frac{2\ atoms}{molecule} \times \frac{6.02 \times 10^{23}\ molecules}{mole}$$
$$= 2.01 \times 10^{23}.$$

$$No.\ atoms\ Hg = 2.0 \times 10^{23}.$$

No. atoms Br $= 150$ m-moles $\times \dfrac{1 \text{ mole}}{1000 \text{ m-moles}} \times \dfrac{2 \text{ atoms}}{\text{molecule}}$

$\times \dfrac{6.02 \times 10^{23} \text{ molecules}}{\text{mole}} = 1.81 \times 10^{23}.$

In a comparison of the values, $\frac{1}{6}$ mole H_2 contains the most atoms.

10. Which contains the greatest number of atoms: 71 g of hydrogen, 90 g of oxygen, 32 g of fluorine, or 1500 g of chlorine? Show why.

Soln:
No. atoms H $= 71$ g $H_2 \times \dfrac{1 \text{ mole}}{2.02 \text{ g}} \times \dfrac{2 \text{ atoms}}{\text{molecule}}$

$\times \dfrac{6.02 \times 10^{23} \text{ molecules}}{\text{mole}} = 4.23 \times 10^{25}.$

No. atoms O $= 90$ g $O_2 \times \dfrac{1 \text{ mole}}{32.0 \text{ g}} \times \dfrac{2 \text{ atoms}}{\text{molecule}}$

$\times \dfrac{6.02 \times 10^{23} \text{ molecules}}{\text{mole}} = 3.39 \times 10^{24}.$

No. atoms F $= 32$ g $F_2 \times \dfrac{1 \text{ mole}}{38.0 \text{ g}} \times \dfrac{2 \text{ atoms}}{\text{molecule}}$

$\times \dfrac{6.02 \times 10^{23} \text{ molecules}}{\text{mole}} = 1.01 \times 10^{24}.$

No. atoms Cl $= 1500$ g $Cl_2 \times \dfrac{1 \text{ mole}}{71.0 \text{ g}} \times \dfrac{2 \text{ atoms}}{\text{molecule}}$

$\times \dfrac{6.02 \times 10^{23} \text{ molecules}}{\text{mole}} = 2.54 \times 10^{25}.$

In a comparison of the values, 71 g of H_2 contains the most atoms.

13. Calculate the weight of one atom of gold from its gram-atomic weight and Avogadro's number.

Soln:
Wt Au $= \dfrac{196.97 \text{ g}}{\text{GAW}} \times \dfrac{1 \text{ GAW}}{6.022 \times 10^{23} \text{ atoms}} = 3.271 \times 10^{-22} \text{ g}$

15. Calculate the percentage composition of each of the following to three significant figures:
(a) C_3H_8.

Soln: $\%C = \dfrac{\text{Wt C}}{\text{FW } C_3H_8} \times 100 = \dfrac{3(12.0)}{3(12.0) + 8(1.01)} \times 100 = 81.7\%$

$\%H = \dfrac{\text{Wt H}}{\text{FW } C_3H_8} \times 100 = \dfrac{8(1.01)}{44.1} \times 100 = 18.3\%$

(b) Na_3PO_4.

Soln: $\%Na = \dfrac{\text{Wt Na}}{\text{FW Na}_3\text{PO}_4} \times 100 = \dfrac{3(23.0)}{164} \times 100 = 42.1\%$

$\%P = \dfrac{\text{Wt P}}{164} \times 100 = \dfrac{31.0}{164} \times 100 = 18.9\%$

$\%O = \dfrac{\text{Wt O}}{164} \times 100 = \dfrac{64.0}{164} \times 100 = 39.0\%$

18. (a) What is the simplest formula of the compound which has the following percentage composition: carbon, 37.51%; hydrogen, 3.15%; fluorine, 59.34%?

Soln: (a) Divide each percentage by the respective atomic weights to obtain a ratio of atoms.

$$C \quad \dfrac{37.51}{12.00} = 3.13$$

$$H \quad \dfrac{3.15}{1.00} = 3.15$$

$$F \quad \dfrac{59.34}{19.00} = 3.12$$

The atomic ratio can be simplified by dividing each value by the smallest value. Therefore, on division the ratio becomes $1 : 1 : 1$, and the simplest formula is CHF.

(b) The molecular weight of the compound is 64. What is its molecular formula?

Soln: (b) Since the molecular formula must be a multiple of the simplest formula, the MW, 64, must be a multiple of the simplest formula weight. The simplest formula, CHF, has a formula weight of $12 + 1 + 19 = 32$. Therefore, $\dfrac{64}{32} = 2$, and the molecular formula of the substance must be $(CHF)_2$, or $C_2H_2F_2$.

19. A 5.00-g sample of an oxide of lead contains 4.33 g of lead. Derive the simplest formula for the compound.

Soln: Calculate the percentage composition of the compound.

$$\%Pb = \dfrac{4.33\ g}{5.00\ g} \times 100 = 86.6\%$$

$$\%O = (100.0 - 86.6)\% = 13.4\%$$

As in Problem 18, divide each percentage by the respective atomic weights to get the atomic ratio.

$$\text{Pb} \quad \frac{86.6}{207.2} = 0.418$$

$$\text{O} \quad \frac{13.4}{16.0} = 0.838$$

Divide by the smallest value to obtain the lowest ratio.

$$\text{Pb} \quad \frac{0.418}{0.418} = 1$$

$$\text{O} \quad \frac{0.838}{0.418} = 2$$

The simplest formula is PbO_2.

25. Determine the simplest formula of a compound which has the following analysis: Na, 16.78%; NH_4, 13.16%; H (except that in NH_4), 0.74%; PO_4, 69.32%.

   Soln: Follow the calculation procedure as in Problems 18 and 19. Divide each percentage by the respective atomic weights or formula weights. Then calculate the simplest ratio.

   $$\text{Na} \quad \frac{16.78}{22.99} = 0.73 \qquad\qquad \text{H} \quad \frac{0.74}{1.01} = 0.73$$

   $$NH_4 \quad \frac{13.16}{18.00} = 0.73 \qquad\qquad PO_4 \quad \frac{69.32}{94.97} = 0.73$$

   Since all values are the same, the ratio is 1 : 1 : 1 : 1, or $Na(NH_4)HPO_4$.

27. Calculate the simplest formula for the compound that has the following composition: Cr, 20.78%; H, 5.64%; C, 48.00%; and O, 25.58%.

   Soln: Follow the same type of calculation procedure as used in Problem 25.

   $$\text{Cr} \quad \frac{20.78}{52.0} = 0.40 \qquad\qquad \text{H} \quad \frac{5.64}{1.01} = 5.58$$

   $$\text{C} \quad \frac{48.00}{12.0} = 4.00 \qquad\qquad \text{O} \quad \frac{25.58}{16.0} = 1.60$$

   Divide by the smallest value to get the smallest ratio.

   $$\text{Cr} \quad \frac{0.40}{0.40} = 1.0 \qquad\qquad \text{H} \quad \frac{5.58}{0.40} = 14.0$$

   $$\text{C} \quad \frac{4.00}{0.40} = 10.0 \qquad\qquad \text{O} \quad \frac{1.60}{0.40} = 4.0$$

   The simplest formula is $CrC_{10}H_{14}O_4$, or, as commonly expressed, $Cr(C_5H_7O_2)_2$.

30. What weight of Al_2O_3 will contain 30.00 g of oxygen?

Soln: First calculate the percentage of oxygen in Al_2O_3.

$$\%O = \frac{3(15.999)}{2(26.98) + 3(15.999)} \times 100 = 47.076\%$$

Then 30.00 g of oxygen is 47.076% of the amount of Al_2O_3 required.

$$47.08\% \ (Wt \ Al_2O_3) = 30.00 \ g$$

$$Wt \ Al_2O_3 = \frac{30.00 \ g}{0.47076} = 63.73 \ g$$

33. A 13.4-millimole sample of a molecular compound weighs 3.45 g. Calculate the molecular weight of the compound.

Soln: A 13.4 m-mole sample is equivalent to 0.0134 mole and is 3.45 g:

$$0.0134 \ mole = 3.45 \ g$$

$$1.0 \ mole = \frac{3.45 \ g}{0.0134 \ mole} = 257 \ g$$

35. What weight of NaN_3 will be needed to produce 100 g of nitrogen (N_2): ($3NaN_3 + \Delta \rightarrow Na_3N + 4N_2$)?

Soln: Calculations involving the weights of substances required to produce a specified amount of product, or the amount of product that can be produced from a given amount of reactant, are necessary prior to carrying out chemical reactions. The calculation requested in this problem is typical. In general, such problems can be resolved in the following manner.

1. Write and balance the chemical equation describing the reaction.
2. Identify the known and unknown entities and label the balanced equation accordingly.
3. Write the mole relation for the components of the reaction below the balanced equation.
4. Solve for the unknown by relating moles of known substance to the ratio of moles of unknown to known and to the conversion factor that converts moles of the unknown to the desired units for the unknown.

   In this problem, 3 moles of NaN_3 produce 4 moles of N_2. Then how many moles of N_2 are in 100 g of N_2, and with a ratio of 3 moles of NaN_3 to 4 moles of N_2, how many moles of NaN_3 are required for the reaction? The weight of NaN_3 is calculated from the number of moles of NaN_3 required. To calculate the number of moles of NaN_3 required for the reaction, calculate the number of moles of N_2 in 100 g of N_2 and then multiply that value by the mole ratio of NaN_3 to N_2. The general procedure is shown herewith:

$$Wt? \qquad\qquad\qquad 100 \text{ g}$$

$$3NaN_3 + \Delta \rightarrow Na_3N + 4N_2$$

$$3 \text{ moles} \qquad \rightarrow 1 \text{ mole} + 4 \text{ moles}$$

GMW values: $NaN_3 = 65.0$ g; $Na_3N = 83.0$ g; $N_2 = 28.0$ g.

$$Wt\ NaN_3 = (100 \text{ g } N_2 \times \frac{1 \text{ mole}}{28.0 \text{ g}}) (\frac{3 \text{ moles } NaN_3}{4 \text{ moles } N_2}) (\frac{65.0 \text{ g } NaN_3}{\text{mole}})$$

$$= 174 \text{ g}$$

36. Calculate the weight of oxygen produced by the reaction of 379 g of potassium nitrate: $(2KNO_3 + \Delta \rightarrow 2KNO_2 + O_2)$.

Soln: $\qquad\qquad$ 379 g $\qquad\qquad$ Wt?

$$2KNO_3 \overset{\Delta}{\rightarrow} 2KNO + O_2$$

$$2 \text{ moles} \rightarrow 2 \text{ moles} + 1 \text{ mole}$$

GMW values: $KNO_3 = 101.0$ g; $O_2 = 32$ g.

$$Wt\ O_2 = (379 \text{ g } KNO_3 \times \frac{1 \text{ mole}}{101.0 \text{ g}} \times \frac{1 \text{ mole } O_2}{2 \text{ moles } KNO_3} \times \frac{32.0 \text{ g}}{\text{mole}} = 60.0 \text{ g}$$

40. Write the balanced equation and calculate the number of moles of carbon monoxide (CO) required to react with 15.0 g of nickel metal (Ni) to produce tetracarbonyl nickel $[Ni(CO)_4]$. How many grams of CO would this be?

Soln: $\qquad$ 15.0 g $\qquad$ No. moles and wt?

$$Ni + 4CO \rightarrow Ni(CO)_4$$

$$1 \text{ mole} + 4 \text{ moles} \rightarrow 1 \text{ mole}$$

GAW Ni = 58.7 g; GMW CO = 28.0 g.

$$\text{No. moles CO} = (15.0 \text{ g Ni} \times \frac{1 \text{ mole}}{58.7 \text{ g}}) (\frac{4 \text{ moles CO}}{1 \text{ mole Ni}}) = 1.02 \text{ mole}$$

$$Wt\ CO = 1.02 \text{ moles} \times \frac{28.0 \text{ g}}{\text{mole}} = 28.6 \text{ g}$$

42. DDT, a chlorinated hydrocarbon, was analyzed. When a 22.21-mg sample was burned (reaction of DDT with oxygen of the air), 38.60 mg of CO_2 and 5.079 mg of H_2O were obtained. What is the simplest formula of DDT? The molecular weight of DDT is approximately 350. What is the molecular formula of DDT?

Soln: GMW values: DDT = 350 g; $CO_2 = 44.0$ g; $H_2O = 18.0$ g.

15

$$\text{Wt C in CO}_2 = 38.60 \text{ mg CO}_2 \times \frac{12.0 \text{ mg C}}{44.0 \text{ mg CO}_2} = 10.53 \text{ mg}$$

$$\text{Wt H in H}_2\text{O} = 5.079 \text{ mg H}_2\text{O} \times \frac{2.0 \text{ mg H}}{18.0 \text{ mg H}_2\text{O}} = 0.564 \text{ mg}$$

Wt Cl = wt sample − wt (C + H)

$$= 22.21 \text{ mg} - 11.09 = 11.12 \text{ mg}$$

Next calculate the ratio of atoms from the data.

$$\text{C} \quad 10.53 \text{ mg} \times \frac{1 \text{ m-mole}}{12 \text{ mg}} = 0.878 \text{ m-mole}$$

$$\text{H} \quad 0.564 \text{ mg} \times \frac{1 \text{ m-mole}}{1.0 \text{ mg}} = 0.564 \text{ m-mole}$$

$$\text{Cl} \quad 11.12 \text{ mg} \times \frac{1 \text{ m-mole}}{35.5 \text{ mg}} = 0.313 \text{ m-mole}$$

Divide each term by 0.313 m-mole to get the smallest ratio,

$$\text{C} \quad \frac{0.878}{0.313} = 2.8; \quad \text{H} \quad \frac{0.564}{0.313} = 1.80$$

$$\text{Cl} \quad = \frac{0.313}{0.313} = 1.0$$

giving $C_{2.8}$, $H_{1.8}$, $Cl_{1.0}$. Multiply by 5 to get the smallest whole number ratio: $C_{14}H_9Cl_5$. Next, determine the molecular formula from the empirical formula and the GMW.

$$(C_{14}H_9Cl_5)_n = \text{molecular formula}$$
$$(168 + 9 + 177.5)_n = 350$$
$$(354)_n = 350; \quad n \simeq 1$$

Therefore, the molecular formula is $C_{14}H_9Cl_5$.

46. What weight of copper(II) nitrate, $Cu(NO_3)_2$, must be decomposed to produce 1500 g (assume the zeroes are significant figures) of copper(II) oxide, CuO? (Work to the nearest hundredth.)

$$Cu(NO_3)_2 \rightarrow CuO + NO_2 + O_2 \text{ (unbalanced)}$$

Soln: The balanced equation is

$$\begin{array}{ccccccc} \text{Wt?} & & 1500 \text{ g} & & & & \\ 2Cu(NO_3)_2 & \rightarrow & 2CuO & + & 4NO_2 & + & O_2 \\ 2 \text{ moles} & \rightarrow & 2 \text{ moles} & + & 4 \text{ moles} & + & 1 \text{ mole} \end{array}$$

GMW values: $Cu(NO_3)_2 = 187.54$ g; CuO = 79.54 g.

$$\text{Wt Cu(NO}_3)_2 = 1500 \text{ g CuO} \times \frac{1 \text{ mole}}{79.54 \text{ g}} \times \frac{2 \text{ moles Cu(NO}_3)_2}{2 \text{ moles CuO}}$$

$$\times \frac{187.54 \text{ g}}{\text{mole}} = 3537 \text{ g}$$

49. What weight of chlorine would be required in the reaction with 10.0 metric tons of hydrogen ($H_2 + Cl_2 \rightarrow 2HCl$)? What weight of HCl would be produced, assuming the yield to be 98.0 percent?

Soln:

$$\begin{array}{cccc} 10 \text{ m-tons} & & \text{Wt?} & \text{Wt?} \\ H_2 & + & Cl_2 & \rightarrow & 2HCl \\ 1 \text{ mole} & + & 1 \text{ mole} & \rightarrow & 2 \text{ moles} \end{array}$$

The mole ratio is valid for all mass or weight measurement units so long as the same unit is used throughout the calculation. In this case, a mole ratio using metric tons for all calculations is most convenient.

Formula weight: H_2 = 2.016 m-tons; Cl_2 = 70.906 m-tons; HCl = 36.5 m-tons.

$$\text{Wt Cl}_2 = 10.0 \text{ m-tons H}_2 \times \frac{70.906 \text{ m-tons Cl}_2}{2.016 \text{ m-tons H}_2} = 352 \text{ m-tons}$$

$$\text{Wt HCl at 100\%} = 10.0 \text{ m-tons H}_2 \times \frac{2(36.461 \text{ m-tons}) \text{ HCl}}{2.016 \text{ m-tons H}_2} = 361.7 \text{ m-tons}$$

Since the HCl yield is only 98 percent of the theoretical yield, the amount of HCl produced is

$$\text{Wt HCl} = (0.98) (361.7 \text{ m-tons}) = 354 \text{ m-tons}$$

50. A solution containing 50.0 g of LiOH is mixed with a solution containing 50.0 g of HCl (LiOH + HCl → LiCl + H_2O). Which chemical is in excess and what weight of LiCl is produced?

Soln: For this type of problem (usually referred to as a limiting reagent type), calculate the number of moles of each reagent available for reaction and according to the mole ratio decide which reagent is in excess.

$$\begin{array}{cccc} 50.0 \text{ g} & 50.0 \text{ g} & \text{Wt?} \\ \text{LiOH} & + \text{ HCl} & \rightarrow & \text{LiCl} & + \text{ H}_2\text{O} \\ 1 \text{ mole} & + 1 \text{ mole} & \rightarrow & 1 \text{ mole} \end{array}$$

GMW values: LiOH = 23.90 g; HCl = 36.5 g; LiCl = 42.4 g.

$$\text{No. moles LiOH} = 50.0 \text{ g} \times \frac{1 \text{ mole}}{23.90 \text{ g}} = 2.09$$

$$\text{No. moles HCl} = 50.0 \text{ g} \times \frac{1 \text{ mole}}{36.5 \text{ g}} = 1.37$$

In a complete reaction of 1.37 moles of HCl with LiOH, 1.37 moles of LiOH will be consumed. Therefore, $2.09 - 1.37$ mole of LiOH will remain in the reaction mixture. The amount of LiCl produced is limited by the amount of HCl available for reaction. Hence, the term *limiting reagent*.

$$\text{Wt LiCl} = 1.37 \text{ moles HCl} \times \frac{1 \text{ mole LiCl}}{1 \text{ mole HCl}} \times \frac{42.4 \text{ g}}{\text{mole}} = 58.1 \text{ g}$$

51. A mixture containing 248.1 g of H_2O_2 and 32.00 g of N_2H_4 reacts as follows: $7H_2O_2 + N_2H_4 \rightarrow 2HNO_3 + 8H_2O$. What is the composition of the system by weight after the reaction, assuming the reaction goes to completion?

Soln: 248.1 g   32.0 g   Wt?   Wt?

$$7H_2O_2 + N_2H_4 \rightarrow 2HNO_3 + 8H_2O$$

7 moles + 1 mole → 2 moles + 8 moles

GMW values: H_2O_2 = 340 g; N_2H_4 = 32.0 g; HNO_3 = 63.0 g; H_2O = 18.0 g.

$$\text{No. moles } H_2O_2 = 248.1 \text{ g} \times \frac{1 \text{ mole}}{34.01 \text{ g}} = 7.295 \text{ moles}$$

$$\text{No. moles } N_2H_4 = 32.00 \text{ g} \times \frac{1 \text{ mole}}{32.04 \text{ g}} = 0.9987 \text{ mole}$$

Decide which is the limiting reagent: 0.9987 mole of N_2H_4 requires 6.991 moles of H_2O_2. However, 7.295 moles of H_2O_2 are available for reaction. Therefore, $7.295 - 6.991$ moles of H_2O_2 remains unreacted.

$$\text{Wt HNO}_3 = 0.9987 \text{ mole } N_2H_4 \times \frac{2 \text{ moles HNO}_3}{1 \text{ mole } N_2H_4} \times \frac{63.0 \text{ g}}{\text{mole}} = 125.8 \text{ g}$$

$$\text{Wt H}_2\text{O} = 0.9987 \text{ mole } N_2H_4 \times \frac{8 \text{ moles H}_2\text{O}}{1 \text{ mole } N_2H_4} \times \frac{18.02 \text{ g}}{\text{mole}} = 143.9 \text{ g}$$

$$\text{Wt H}_2\text{O}_2 \text{ remaining} = 0.304 \text{ mole} \times \frac{34.0 \text{ g}}{\text{mole}} = 10.3 \text{ g}$$

53. What weight of $P(CH_3)_4I$ will be formed by the reaction of 16.0 lb of CH_3I with $P(CH_3)_3$? ($P(CH_3)_3 + CH_3I \rightarrow P(CH_3)_4I$.) What weight of impure $P(CH_3)_3$ (90 percent pure by weight) will be required in this reaction?

Soln: Wt?   16.0 lb   Wt?

$$P(CH_3)_3 + CH_3I \rightarrow P(CH_3)_4I$$

1 mole + 1 mole → 1 mole

Formula weight values: $P(CH_3)_3$ = 76.0 lb; CH_3I = 142.0 lb; $P(CH_3)_4I$ = 218.0 lb.

$$\text{Wt } P(CH_3)_4 I = 16.0 \text{ lb } CH_3 I \times \frac{218.0 \text{ lb } P(CH_3)_4 I}{142.0 \text{ lb } CH_3 I} = 24.6 \text{ lb}$$

$$\text{Wt } P(CH_3)_3 \text{ (pure)} = 16.0 \text{ lb } CH_3 I \times \frac{76.0 \text{ lb } P(CH_3)_3}{142.0 \text{ lb } CH_3 I} = 8.56 \text{ lb}$$

$$\text{Wt of 90\% pure } P(CH_3)_3 = (0.90) \text{ (Wt of impure } P(CH_3)_3 = 8.56 \text{ lb)}$$

$$= \frac{8.56}{0.90} = 9.5 \text{ lb}$$

UNIT III

Structure of the Atom and the Periodic Law

3

INTRODUCTION

The historical foundations are laid in Chapter 3 for the development of modern atomic theory. The rudimentary discoveries of the last century (the electron, proton, radioactivity, and so on) are traced to the development of the Bohr atom. A self-consistent picture of atomic structure using quantum numbers is then organized drawing on Hund's rule and the Pauli exclusion principle of modern quantum mechanics.

Due to the largely theoretical nature of this chapter, inclusion of several answers to the Questions section was deemed advisable. The answers provided are for questions that have some quantitative involvement.

FORMULAS AND DEFINITIONS

The relation between the frequencies of emitted light and the occupancy of quantum levels in an atom is given by the Rydberg formula

$$\nu = 3.289 \times 10^{15} \left(\frac{1}{n_1{}^2} - \frac{1}{n_2{}^2} \right)$$

where

ν = frequency;

n_1 = an integer representing the energy level closer to the nucleus;

n_2 = level farther from the nucleus.

The energy E_n that an electron has, in ergs, is

$$E_n = \frac{-21.79 \times 10^{-12} \text{ erg}}{n^2}$$

where n is an integer corresponding to the energy level. The energy difference between two levels is calculated from the Bohr equation,

$$E = E_{n_2} - E_{n_1} = 21.79 \times 10^{-12} \frac{\text{erg}}{\text{atom}} \left(\frac{1}{n_1{}^2} - \frac{1}{n_2{}^2} \right)$$

QUESTIONS

20. (a) Why is 12.011 listed in the atomic weight table as the atomic weight of carbon, whereas the arbitrary standard for the atomic weight scale is the exact number 12 for the mass of the carbon-12 isotope?

Soln: The number 12.011 represents the atomic weight of carbon, a weighted average of all isotopes of carbon as they exist in nature. The exact number 12 is the atomic mass of the carbon-12 isotope.

(b) Carbon, as it occurs in nature, contains approximately 99 percent $^{12}_{6}C$, 1 percent $^{13}_{6}C$, and an infinitesimal trace of $^{14}_{6}C$. Using the recognized standard that specifies the exact number 12 amu for the mass of $^{12}_{6}C$, calculate the average atomic weight for the mixture and compare this with the atomic weight listed for carbon in the atomic weight table.

Soln:

$$12 \times 0.99 \ = \ 11.88$$
$$13 \times 0.01 \ = \ \ 0.13$$
$$\text{Total } 100\% \ = \ 12.01$$

This is compared with 12.011 amu, the average atomic weight listed in the table.

28. Why should less energy be required in the removal of an electron from an L electron shell than from a K shell?

Soln: Less energy is required in the removal of an electron from an L electron shell than from a K shell because the K electrons are held more tightly, thus requiring more energy to be removed. (That is, K electrons are at a lower energy relative to the zero of energy than the L electrons are.)

35. How many electrons would be required to balance a single proton on an analytical balance sensitive enough to weigh electrons? How many electrons would be needed to balance a neutron? How many electrons would be needed to weigh one gram?

Soln: On the atomic weight scale, the mass of the electron is 0.00055; the mass of the proton is 1.0073. To balance the mass of the proton, 1.0073 amu/0.00055 amu = 1831.45 electrons would be required. For one gram to be weighed,

$$1.0000 \text{ g} \times \frac{1 \text{ electron}}{0.00055 \text{ amu}} \times \frac{1 \text{ amu}}{1.660531 \times 10^{-24} \text{ g}} = 1.0949 \times 10^{27} \text{ electrons}$$

would be needed.

50. How does the mass of a neutron compare with the sum of the masses of a proton and an electron? If a neutron should disintegrate, producing a proton and an electron, can all the mass be accounted for?

Soln:

	amu
Mass of neutron	1.0087
Mass of proton	1.0073
Mass of electron	0.00055

sum = 1.0079

The difference in mass of the neutron mass minus the proton mass plus the electron mass is 0.0008 amu. This mass difference is the energy change that Einstein's expression $E = mc^2$ predicts and is responsible for the energy holding the neutron together.

PROBLEMS

1. There are three naturally occurring isotopes of silicon. Their atomic weights are 27.97693, 28.97650, and 29.97377. They constitute 92.210%, 4.7000% and 3.0900% of naturally occurring silicon, respectively. Calculate the atomic weight of naturally occurring silicon (as it is given in the Periodic Table).

   Soln: The atomic weight of naturally occurring silicon equals the contribution of each of the isotopes.

   AW of Si = 92.210% × 27.97693 + 4.7000% × 28.97650 + 3.0900% × 29.97377

   = 25.7975 + 1.3619 + 0.9262

   = 28.0856

   The accepted value is 28.086.

2. Naturally occurring boron consists of two isotopes whose atomic weights are 10.01294 and 11.00931. From these data and the atomic weight of boron given in the Periodic Table, calculate the percentage of each isotope in naturally occurring boron.

Soln: As in Problem 1, the atomic weight must be made up of the correct percentages of the two isotopes. Let x = fraction of 10.01294 isotope, $1 - x$ = fraction of 11.00931 isotope.

$$10.81 = x(10.01294) + (1 - x)(11.00931)$$

$$10.81 - 11.00931 = 10.01294x - 11.00931x$$

$$0.19931 = 0.99637x$$

$$x = 0.2000$$

or 20.00% and 80.00% respectively.

3. Compute the frequency of radiation emitted for the electron transitions between the following energy levels in the hydrogen atom: (a) from $n = 2$ to $n = 1$ (i.e., $n_1 = 1$ and $n_2 = 2$).

Soln: (a) $\nu = 3.289 \times 10^{15} \left(\dfrac{1}{n_1{}^2} - \dfrac{1}{n_2{}^2}\right) \sec^{-1}$

$\qquad = 3.289 \times 10^{15} \left(\dfrac{1}{1^2} - \dfrac{1}{2^2}\right)$

$\qquad = (3.289 \times 10^{15})0.75$

$\qquad = 2.467 \times 10^{15}$ cycles per sec

4. Calculate the amount of energy emitted for the electron transitions in Problem 3.

Soln: (a) $E = 21.79 \times 10^{-12}$ erg/atom $\left(\dfrac{1}{n_1{}^2} - \dfrac{1}{n_2{}^2}\right)$

$\qquad = 21.79 \times 10^{-12} (0.75)$

$\qquad = 1.634 \times 10^{-11}$ erg/atom

UNIT IV

The Gaseous State and the Kinetic-Molecular Theory

10

INTRODUCTION

This chapter treats concepts associated with the kinetic-molecular theory. The effects on a confined gas system produced by changes in pressure, volume, temperature, or the number of moles are treated in detail. Also, the laws of Gay-Lussac and Avogadro describing reactions are presented along with ramifications resulting from the involvement of the variables mentioned above in the reaction system.

The problems in this chapter represent those types of calculations involving the properties unique to gases and interpretation of data derived from experiments that include gases as reactants or products.

FORMULAS AND DEFINITIONS

Gas pressure The pressure of a gas is normally measured in the laboratory by determining the length of a column of mercury that the gas can support. The unit of pressure used in this textbook is the mmHg or the cmHg. The conversion relation of mmHg to atmosphere (atm) is 760 mmHg = 1.0 atm.

S.T.P. This abbreviation applies to studies involving gases and is read "standard temperature and pressure." The standard temperature is 0°C, or 273°K, and 1.0 atm, or 760 mmHg pressure.

Boyle's law The volume of a confined gas held at constant temperature is inversely related to the applied pressure. The algebraic representation is $P_1 V_1 = P_2 V_2$.

Charles's law   The volume of a confined gas held at constant pressure is directly related to the temperature (°K) of the gas. The algebraic representation is, $V_1/V_2 = T_1/T_2$.

Boyle's law and Charles's law combined   By appropriate algebraic manipulations, P, V, and T can be related as

$$\frac{P_1 V_1}{T_1} = \frac{P_2 V_2}{T_2}$$

Gas constant (R)   Calculated from the ideal gas equation of state, $PV = nRT$. The value of R is 0.082 liter · atm/mole · °K.

Ideal gas equation, $PV = nRT$ The equation relates the variables volume, pressure, number of moles, and temperature of a confined gas. In the equation, the units for the variables are P in atmospheres, V in liters, n in moles, and T in °K.

Molar volume   The volume of one mole of gas measured at 1.0 atm of pressure; at 273°K, it is approximately 22.4 liters.

Partial pressure   The total pressure a mixture of gases exerts is the sum of the pressures that each gas exerts.

$$P_{\text{total}} = p_1 + p_2 + p_3 + \cdots + p_n$$

PROBLEMS

1. The volume of a mass of gas is 4.23 liters at 747 mm and 24°C. What volume will it occupy at 700 mm and 24°C?

   Soln: Problems involving gases usually can be solved in a logical fashion by analyzing the problem in terms of the variables. A change in any of the factors V, T, P, or n (moles of gas) for a confined gas will cause a change in the gas system. To begin, make a table like the following, including these factors:

   | V_1 | T_1 | P_1 | n_1 |
   | V_2 | T_2 | P_2 | n_2 |

   The subscript 1 is used to denote the initial condition of the gas; the subscript 2 denotes the new or final condition of the gas. For this problem, insert the values as shown:

   | V_1 4.23 liters | T_1 24°C, or 297°K | P_1 747 mm | n_1 |
   | V_2 ? | T_2 24°C, or 297°K | P_2 700 mm | n_2 |

   When the data are examined, it is evident that T is held constant and the mass or number of moles of gas is constant. The pressure

25

on the gas is being decreased from 747 mm to 700 mm. The volume will change by the same proportion as the change in pressure, in inverse relation. In other words, a decrease in pressure will cause an increase in volume, or volume is inversely proportional to pressure:

$$V_2 = 4.23 \text{ liters} \times \frac{747 \text{ mm}}{700 \text{ mm}} = 4.51 \text{ liters}$$

3. The volume of a given mass of gas is 410 ml at 33.0°C and 467 mm. What will its volume be if it is measured at 10.0°C and 467 mm?

Soln: Analyze the problem as in Problem 1.

V_1	410 ml	T_1	33°C = 306°K	P_1	467 mm	n_1
V_2	?	T_2	10°C = 283°K	P_2	467 mm	n_2

Observe that in this case n and P are held constant, and that the temperature is decreasing from 306°K to 283°K. The volume will change by the same proportion as the temperature and in the same direction. The temperature is decreased by the proportion 283°K/306°K, so V_2 must equal

$$V_2 = 410 \text{ ml} \times \frac{283°K}{306°K} = 379 \text{ ml}$$

5. The volume of a gas is 800 ml at S.T.P. What volume will it occupy at 55°C and 790 mm?

Soln: S.T.P. conditions are 0°C and 760 mmHg.

V_1	800 ml	T_1	0°C = 273°K	P_1	760 mm	n_1
V_2	?	T_2	55°C = 328°K	P_2	790 mm	n_2

The volume is directly proportional to an increase in T and inversely proportional to an increase in P. The number of moles is held constant. The changes can be considered in the following manner. First, the increase in T produces an increase in V by the proportion 328°K/273°K. Second, the increase in P produces a decrease in V by the proportion 760 mm/790 mm. Collectively, the changes can be written

$$V_2 = 800 \text{ ml} \times \frac{328°K}{273°K} \times \frac{760 \text{ mm}}{790 \text{ mm}} = 925 \text{ ml}$$

7. A gas occupies 300 ml at −10° C and 720 mm. What pressure will the gas exert in a 505-ml sealed bulb at 25°C?

Soln:

V_1	300 ml	T_1	−10°C = 263°K	P_1	720 mm	n_1
V_2	505 ml	T_2	25°C = 298°K	P_2	?	n_2

26

In a sealed container, the number of moles of gas cannot change. The temperature is increased by the proportion $298°K/263°K$, which is a factor of 1.13; but the volume is changed by a factor of 505 ml/300 ml, or 1.68. Therefore, the pressure must have decreased to effect the change. The data are most easily treated by solving the combined equation for P_2 and making the substitutions:

$$\frac{P_1 V_1}{T_1} = \frac{P_2 V_2}{T_2} \qquad \text{or} \qquad P_2 = \frac{V_1 P_1 T_2}{V_2 T_1}$$

$$P_2 = \frac{(300 \text{ ml})(720 \text{ mm}) (298°K)}{(505 \text{ ml}) (263°K)} = 485 \text{ mm}$$

9. The volume of gas collected over water at 32.0°C and 752 mm is 627 ml. What will the volume of the gas be when dried and measured at S.T.P.? (Vapor pressure of water at 32°C = 35.7 mm.)

Soln: The gas is collected over water. The total pressure, 752 mm, is the pressure exerted by the gas plus the pressure exerted by the water vapor. Therefore the pressure of the gas is

$P_{total} = P_{gas} + P_{H_2O}$

$P_{gas} = P_{total} - P_{H_2O} = 752 \text{ mm} - 35.7 \text{ mm} = 716 \text{ mm}$

V_1 627 ml T_1 32.0°C = 305°K P_1 716 mm

V_2 ? T_2 0°C = 273°K P_2 760 mm

$$V_2 = 627 \text{ ml} \times \frac{273°K}{305°K} \times \frac{716 \text{ mm}}{760 \text{ mm}} = 529 \text{ ml}$$

12. Calculate the relative rates of diffusion of the gases CO and Ne.

Soln: In comparing the rates of diffusion of two gases, it is important to understand that the gas with the higher molecular weight will diffuse at a slower rate than the gas with the lower molecular weight. Also, the higher-molecular-weight molecules travel at a lower speed than the less massive molecules. The rates are compared by applying Graham's law:

$$\frac{R_1}{R_2} = \sqrt{\frac{M_2}{M_1}}$$

When using the equation, one can simplify the arithmetic by letting M_2 and R_2 represent the more massive molecule.
   The formula weights of the gases are CO = 28.0, Ne = 20.2.

$$\frac{R_{Ne}}{R_{CO}} = \sqrt{\frac{28.0}{20.2}} = \sqrt{1.39} = 1.18$$

$$R_{Ne} = 1.18 \, R_{CO}$$

Neon diffuses 1.18 times faster than CO.

17. A gas of unknown composition diffuses at the rate of 10 ml/sec in a diffusion apparatus in which CH_4 gas diffuses at the rate of 30 ml/sec. Calculate the approximate molecular weight of the gas of unknown composition.

Soln: The unknown gas diffuses at a slower rate than CH_4. Therefore, the unknown gas has a molecular weight greater than CH_4:

$$\frac{R_{CH_4}}{R_{unknown}} = \sqrt{\frac{MW_{unknown}}{MW_{CH_4}}} \quad \text{or} \quad \frac{30 \text{ ml/sec}}{10 \text{ ml/sec}} = \sqrt{\frac{MW}{16}}$$

Squaring both sides of the equation and division yields

$$3^2 = \frac{MW}{16} \quad \text{or} \quad MW = (9)(16) = 144$$

Only two significant figures are allowed; hence, 144 is rounded to 140.

18. Calculate the volume occupied by the oxygen produced by the thermal decomposition of 75.0 g of $KClO_3$, assuming the oxygen to be measured at 20.0°C and 743 mm. (Density of oxygen at S.T.P. is 1.429 g/liter.)

Soln: This problem involves a chemical reaction in which a gas is generated as a product. The volume of gas formed in a reaction is calculated by recognizing that one mole of any gas measured at S.T.P. occupies a volume of approximately 22.4 liters. Make the calculations as though the reaction occurs at S.T.P. conditions and then make any necessary changes if the conditions are different from those at S.T.P.

$$\begin{array}{cc} 75.0 \text{ g} & \text{Wt?} \\ 2KClO_3 \rightarrow 2KCl + 3O_2 \\ 2 \text{ moles} \rightarrow & 3 \text{ moles} \end{array}$$

$$\text{Wt } O_2 = (75.0 \text{ g } KClO_3 \times \frac{1 \text{ mole}}{122.5 \text{ g}}) (\frac{3 \text{ moles } O_2}{2 \text{ moles } KClO_3}) (\frac{32 \text{ g } O_2}{\text{mole}}) = 29.4 \text{ g}$$

$$V_{O_2} \text{ at S.T.P.} = 29.4 \text{ g} \times \frac{1.0 \text{ liter}}{1.429 \text{ g}} = 20.6 \text{ liters}$$

At 20.0°C and 743 mm, the volume will be

$$V = 20.6 \text{ liters} \times \frac{760 \text{ mm}}{743 \text{ mm}} \times \frac{293°K}{273°K} = 22.6 \text{ liters}$$

22. Calculate the weight of dry hydrogen in 750 ml of moist hydrogen gas collected over water at 25.0°C and 755 mm. (The density of hydrogen is 0.08987 g/liter at S.T.P.) (See Table 10-1.)

Soln: Since the hydrogen was collected over water, the total volume, 750 ml, is composed of water vapor and hydrogen. The volume of hydrogen is proportional to the partial pressure of hydrogen. p_{H_2O} at $25°C = 23.8$ mm; and $p_{H_2} = 755$ mm $- 23.8$ mm $= 731$ mm.

V_1 750 ml	T_1 $25°C = 298°K$	P_1 731 mm
V_2 ?	T_2 $273°K$	P_2 760 mm

$$V \text{ at S.T.P.} = 750 \text{ ml} \times \frac{273°K}{298°K} \times \frac{731 \text{ mm}}{760 \text{ mm}} = 661 \text{ ml}$$

Wt $H_2 = 0.661$ liter $\times 0.0899$ g/liter $= 0.0594$ g

24. Air at S.T.P. has a density of 1.292 g/l. What will be the weight of 4.67 liters of air at 90°C and 735 mm?

Soln: Convert the volume to that measured at S.T.P. conditions and then multiply the volume and the density at S.T.P.:

$$V \text{ at S.T.P.} = 4.67 \text{ liters} \times \frac{273°K}{363°K} \times \frac{735 \text{ mm}}{760 \text{ mm}} = 3.40 \text{ liters}$$

Wt $= 3.40$ liters $\times 1.29$ g/liter $= 4.39$ g

30. Calculate the partial pressures of oxygen and of nitrogen in the air at a total pressure of 760 mm for air that is 20.8 percent oxygen and 79.2 percent nitrogen by volume.

Soln: The pressure of each gas is proportional to its volume fraction, or percentage.

$$p_{N_2} = (0.792) (760 \text{ mm}) = 602 \text{ mm}$$
$$p_{O_2} = (0.208) (760 \text{ mm}) = 158 \text{ mm}$$

33. The time of outflow of a gas through a small opening is 32.6 minutes, while that of an equal number of moles of hydrogen is 5.50 minutes. Calculate the molecular weight of the first gas.

Soln: To use the diffusion relation, recognize that rate in this instance is a quantity of gas divided by time. The unknown gas diffuses at a lower rate than hydrogen and must have a higher molecular weight than hydrogen.

$$\frac{R_{H_2}}{R_{unknown}} = \sqrt{\frac{MW}{MW_{H_2}}} = \text{ or } \frac{\text{moles } H_2/5.50 \text{ min}}{\text{moles unknown}/32.6 \text{ min}} = \sqrt{\frac{MW}{2(1.008)}}$$

Since moles H_2 = moles unknown, the equation reduces to

$$\frac{32.6}{5.5} = \sqrt{\frac{MW}{2.016}}$$

Squaring both sides of the equation and multiplication yields

$$MW = 2.016 \, (\frac{32.6}{5.50})^2 = 70.8$$

35. (a) When two cotton plugs, one moistened with ammonia and the other with hydrochloric acid, are simultaneously inserted into opposite ends of a glass tube 97.0 cm long, a white ring of NH_4Cl forms where gaseous NH_3 and gaseous HCl first come into contact $(NH_3 + HCl \rightarrow NH_4Cl)$. At what distance from the ammonia-moistened plug does this occur?

Soln: The distance of the ring from the NH_3 end is proportional to the rates at which NH_3 and HCl diffuse.

$$\frac{R_{NH_3}}{R_{HCl}} = \sqrt{\frac{MW \ HCl}{MW \ NH_3}}$$

$$\frac{R_{NH_3}}{R_{HCl}} = \sqrt{\frac{36.46}{17.03}} = \sqrt{2.14} = 1.46$$

$$R_{NH_3} = 1.46 \, R_{HCl}$$

Distance NH_3 + distance HCl = 97.0 cm.

$$1.46 \, D_{HCl} + D_{HCl} = 97.0 \text{ cm}$$

$$2.46 \, D_{HCl} = 97.0 \text{ cm}$$

$$D_{HCl} = 39.4 \text{ cm}$$

$$D_{NH_3} = (97.0 - 39.4) \text{ cm} = 57.6 \text{ cm.}$$

36. What volume is occupied by 40.0 g of fluorine (a) At S.T.P.? (b) At 25.0° and 900 mm?

Soln: (a) The volume is calculated from the molar volume relation at S.T.P.: 22.4 liters ≡ 1 mole gas.

$$GMW_{F_2} = 38.0 \text{ g}$$

$$V_{F_2} \text{ at S.T.P.} = 40.0 \text{ g} \times \frac{1 \text{ mole}}{38.0 \text{ g}} \times \frac{22.4 \text{ liters}}{\text{mole}} = 23.6 \text{ liters}$$

(b) $V = 23.6 \text{ liters} \times \frac{760 \text{ mm}}{900 \text{ mm}} \times \frac{298°K}{273°K} = 21.7 \text{ liters}$

40. How many moles are represented by each of the following, measured at S.T.P.? (a) 3.00 liters of acetylene, C_2H_2 (b) 50 ml of ammonia, NH_3

Soln: The molar volume relation applies to all parts of the problem.

(a) No. moles = 3.00 liters $C_2H_2 \times \frac{1 \text{ mole}}{22.4 \text{ liters}} = 0.134 \text{ mole}$

(b) No. moles = 50 ml NH_3 × $\dfrac{1.0 \text{ liter}}{1000 \text{ ml}}$ × $\dfrac{1 \text{ mole}}{22.4 \text{ liters}}$

$$= 2.2 \times 10^{-3} \text{ mole}$$

46. What pressure is exerted by 11.0 g of CO when the gas is contained in a 20.0-liter vessel at 27.0°?

Soln: Problems of this kind are most easily solved by applying the ideal gas equation to the data.

$$n_{CO} = 11.0 \text{ g} \times \frac{1 \text{ mole}}{28.01 \text{ g}} = 0.393 \text{ mole}$$

$P = ?; V = 20.0 \text{ liters}; T = 27.0°C = 300°K.$

$PV = nRT \text{ or } P = \dfrac{nRT}{V}.$

$$P = \frac{(0.393 \text{ mole}) (0.082 \text{ liter} \cdot \text{atm/mole} \cdot °K) (300°K)}{20.0 \text{ liters}}$$

$$= 0.483 \text{ atm}$$

48. Calculate the density of each of the following gases at S.T.P. and at 20.0° and 720 mm: NO. . . .

Soln: At S.T.P., a GMW of a substance occupies 22.4 liters. Therefore, the density of NO at S.T.P. is

$$D = \frac{30.01 \text{ g}}{\text{mole}} \times \frac{1 \text{ mole}}{22.4 \text{ liters}} = \frac{1.34 \text{ g}}{\text{liter}}$$

The molar volume at 20°C and 720 mm is expressed

$$V = 22.4 \text{ liters} \times \frac{760 \text{ mm}}{720 \text{ mm}} \times \frac{293°K}{273°K} = 25.38 \text{ liters}$$

$$D = \frac{30.01 \text{ g}}{\text{mole}} \times \frac{1 \text{ mole}}{25.38 \text{ liters}} = \frac{1.18 \text{ g}}{\text{liter}}$$

50. How many liters of dry hydrogen gas, measured at 27.0° and 778 mm, can be obtained by the reaction of 47.0 g of aluminum with an excess of dilute sulfuric acid?

Soln: This problem is similar to Problem 18, but will be solved by taking advantage of the molar volume relation:

$$\begin{array}{cccc} 47.0 \text{ g} & & & V \text{ at S.T.P.} \\ 2Al & + 3H_2SO_4 & \rightarrow Al_2(SO_4)_3 + & 3H_2 \\ 2 \text{ moles} & & \rightarrow & 3 \text{ moles} \end{array}$$

$V_{H_2} = (47.0 \text{ g Al} \times \dfrac{1 \text{ mole}}{27.0 \text{ g}}) (\dfrac{3 \text{ moles } H_2}{2 \text{ moles Al}}) (\dfrac{22.4 \text{ liters}}{\text{mole}})$

$$= 58.49 \text{ liters at S.T.P.}$$

31

The volume at the experimental conditions is expressed as

$$V = 58.49 \text{ liters } \times \frac{760 \text{ mm}}{778 \text{ mm}} \times \frac{300°K}{273°K} = 62.8 \text{ liters}$$

51. A mixture of 0.100 g of hydrogen, 0.500 g of nitrogen, and 0.410 g of argon is stored at S.T.P. What volume must the container have, assuming no interaction of the three gases?

Soln: The volume of the gas mixture is independent of the kinds of gas in the system at 1.0 atm and 273°K, but is dependent on the total number of moles of gas in the system. Applying the ideal gas equation gives

$$V = \frac{nRT}{P}; \qquad n_{total} = n_{H_2} + n_{N_2} + n_{Ar}$$

$$n_{H_2} = 0.100 \text{ g} \times \frac{1 \text{ mole}}{2(1.008 \text{ g})} = 0.0496 \text{ mole}$$

$$n_{Ar} = 0.410 \text{ g} \times \frac{1 \text{ mole}}{39.95 \text{ g}} = 0.0103 \text{ mole}$$

$$n_{N_2} = 0.500 \text{ g} \times \frac{1 \text{ mole}}{2(14.01 \text{ g})} = 0.0178 \text{ mole}$$

$$n_{total} = 0.0777 \text{ mole}$$

$$V = \frac{(0.0777 \text{ mole}) (0.082 \text{ liter} \cdot \text{atm/mole} \cdot °K) (273°K)}{1.0 \text{ atm}}$$

$$= 1.74 \text{ liter}$$

56. Calculate the weight of ethane, C_2H_6, required to produce a pressure of 1520 mm at 20.0° when contained in a 9.00-liter vessel.

Soln: The weight of C_2H_6 required can be determined by calculating first the number of moles of C_2H_6. As an alternative, the ideal gas equation can be rearranged in the following manner to yield the weight of C_2H_6 directly: $PV = nRT$, where $n = Wt/GMW$; GMW $C_2H_6 = 30.07$ g.

Substitution yields $PV = (Wt/GMW)RT$. Solving for Wt gives

$$Wt = \frac{(PV) (GMW)}{RT}$$

$$Wt\ C_2H_6 = \frac{(1520 \text{ mm} \times (1 \text{ atm}/760 \text{ mm}))(9.00 \text{ liter}) (30.07 \text{ g})}{(0.082 \text{ liter} \cdot \text{atm/mole} \cdot °K) (293°K)}$$

$$= 22.5 \text{ g}$$

65. The weight of 151 ml of a certain vapor at 68.0° and 768 mm is 1.384 g. What is the molecular weight of the vapor?

Soln: The version of the ideal gas equation used in Problem 56 is applicable to these data: $PV = (Wt/GMW)RT$, or $GMW = (Wt)(RT)/PV$.

$$P = 768 \text{ mm} \times \frac{1 \text{ atm}}{760 \text{ mm}} = 1.01 \text{ atm}; \qquad V = 0.151 \text{ liter}$$

$$GMW = \frac{(1.384 \text{ g})(0.082 \text{ liter} \cdot \text{atm/mole} \cdot °K)(341°K)}{(1.01 \text{ atm})(0.151 \text{ liter})}$$

$$= 254 \text{ g}$$

68. What weight of MnO_2 is required to produce 2500 liters of Cl_2 gas at 760 mm pressure and 280°C, according to the following reaction: $MnO_2 + 4HCl \rightarrow MnCl_2 + Cl_2 + 2H_2O$?

Soln: First calculate the number of moles of Cl_2 that would be produced at S.T.P.

V_1 2500 liters	P_1 760 mm	T_1 280°C = 553°K
$V_{S.T.P.}$?	P_2 760 mm	T_2 273°K

$$V_{S.T.P.} = 2500 \text{ liters} \times \frac{273°K}{553°K} = 1234 \text{ liters}$$

$$\text{No. moles} = 1234 \text{ liters} \times \frac{1 \text{ mole}}{22.4 \text{ liters}} = 55.1$$

$$MnO_2 + 4HCl \rightarrow MnCl_2 + Cl_2 + 2H_2O$$

$$1 \text{ mole} \qquad \rightarrow \qquad 1 \text{ mole}$$

According to the equation, the weight of MnO_2 is

$$\text{Wt } MnO_2 = 55.1 \text{ moles } Cl_2 \times \frac{1 \text{ mole } MnO_2}{1 \text{ mole } Cl_2} \times \frac{86.9 \text{ g}}{\text{mole}} = 4788 \text{ g}$$

Three significant figures are allowed: 4790 g.

69. Calculate the pressure exerted by a mole of ethylene, C_2H_4, in 10 liters at 100° C using (a) the ideal gas law; (b) the van der Waals equation. For ethylene, the constant a has a value of 4.471 liter² · atm/mole², and the constant b a value of 0.05714 liter/mole.

Soln: This problem is included to show that real gases deviate slightly from ideal behavior.

(a) $PV = nRT$, or $P = nRT/V$.

$$P = \frac{(1 \text{ mole})(0.082 \text{ l} \cdot \text{atm/mole} \cdot °K)(373°K)}{10.0 \text{ liter}}$$

$$= 3.06 \text{ atm}$$

(b) $(P + \frac{a}{V^2})(V - b) = nRT$

Solve for P:

$$P = \frac{nRT}{V-b} - \frac{a}{V^2}$$

$$= \frac{(1 \text{ mole}) (0.082 \text{ liter} \cdot \text{atm/mole} \cdot {}^{\circ}\text{K}) (373{}^{\circ}\text{K})}{(10.0 \text{ liter} - 0.05714 \text{ liter})} -$$

$$\frac{4.471 \text{ liter}^2 \cdot \text{atm/mole}^2}{(10.0 \text{ liter})^2}$$

$$= (\frac{30.59}{9.94} - \frac{4.471}{100}) \text{ atm}$$

$$= 3.077 - 0.0447 = 3.03 \text{ atm}$$

UNIT V

The Liquid
and Solid States
11

INTRODUCTION

Various aspects of the liquid and the solid states are the subject of this chapter. Wherever possible, the interparticle attractive or repulsive forces are used to explain such phenomena as evaporation and condensation, boiling and freezing points, and the heat associated with the transformations between solid, liquid, and gaseous states. Much of the chapter details the structures possible in the solid state. Radius ratio rules are discussed in addition to the Born-Haber Cycle.

FORMULAS AND DEFINITIONS

Heat of vaporization   The quantity of heat necessary to evaporate a unit mass of liquid at a constant temperature. The heat of vaporization of water is 540 calories per gram.

Heat of fusion   The quantity of heat necessary to cause a change from the solid to the liquid state at constant temperature for a unit mass of substance. The heat of fusion of water is approximately 80 calories per gram.

Radius ratio   Radius of cation to radius of anion, that is, r^+/r^-.

Born-Haber Cycle   A means for determining the lattice energy of an ionic crystal using available thermochemical quantities.

Bragg equation   A relation that defines the angle of reflected X-rays as they pass through a crystal. That is, $n\lambda = 2d \sin \theta$, where n = an integer number, λ = the wavelength of X-rays, d = the interatomic distance, and θ = the angle of "grazing incidence."

35

PROBLEMS

Note: Pertinent thermal data for water is given on page 221 of the textbook.

1. If 270 g of steam at 100° and 950 g of ice at 0° are combined, what is the temperature of the resultant water (assume no heat loss)?

   Soln: A heat balance like that used in Chapter 1 is used taking into account the heat changes in the transformations.

   $$Q_{lost} = Q_{gained}$$

   $$Q_{steam} + Q_{H_2O} \text{ (from steam)} = Q_{ice} + Q_{H_2O} \text{ (from ice)}$$

   $$Q_{lost} = 270 \text{ g} \times 540 \text{ cal/g} + 270 \text{ g} \times 1 \text{ cal/g°C} (100 - T_f)$$

   $$Q_{gained} = 950 \text{ g} \times 79.7 \text{ cal/g} + 950 \text{ g} \times 1 \text{ cal/g°C} (T_f - 0)$$

   $$145,800 + 270 (100 - T_f) = 75,715 + 950 \times (T_f - 0)$$

   $$1220 T_f = 97,085$$

   $$T_f = 79.6°C$$

2. How much heat is required to convert 150 g of ice at −40.0° to steam at 135°? The specific heat of ice is 0.50 cal per gram per degree; of water, 1.0; and of steam, 0.48.

   Soln: As in Problem 1, two transformations are occurring as well as the simple heating of the three phases — ice, water, and steam.

Ice region	$Q = 150 \text{ g} \times 80 \text{ cal/g} + 150 \text{ g} (0.50 \text{ cal/g°C}) 40°$
Water region	$+ 150 \text{ g} (1.00 \text{ cal/g°C}) 100°$
Steam region	$+ 150 \text{ g} \times 540 \text{ cal/g} + 150 \text{ g} (0.48 \text{ cal/g°C}) 35°$

   $$Q = 12,000 + 3000 + 15,000 + 81,000 + 2520 = 113,500 \text{ cal}$$

   Rounded to significant figures, 110 kcal.

5. What weight of oxygen must be burned in an excess of hydrogen (heat of formation of water at 25° is 68,315 cal/mole) to furnish a quantity of heat necessary to convert 500 g of ice at 0°? The heat of fusion of water is 80 cal/g at 0°.

   Soln: The heat of formation of water is used to melt ice. Heat required is 500 g × 80 cal/g = 40,000 cal. This heat is supplied by a certain number of grams of water, the oxygen content of which is to be calculated.

   $$40,000 \text{ cal} = \frac{68,315 \text{ cal}}{\text{mole}} \times \frac{1 \text{ mole}}{18 \text{ g H}_2\text{O}} \times \frac{18 \text{ g H}_2\text{O}}{16 \text{ g O}_2} \times x \text{ g O}_2$$

   $$40,000 \text{ cal} = 4269 \text{ cal} \times x \text{ g O}_2$$

   $$\text{Wt O}_2 = 9.3 \text{ g}$$

8. What is the maximum amount of liquid water at 25.0° which could be frozen by the cooling action of vaporizing 50.0 g of liquid ammonia?

Soln: The amount of heat abstracted by the ammonia vaporization equals the amount of heat removed from water cooling from 25°C to 0° and the heat used to transform the water to ice:

$$50.0 \text{ g} \times 327 \text{ cal/g} = x \text{ g} (1.0 \text{ cal/g°C}) (25°C) + 80 \text{ cal/g} (x \text{ g})$$

$$16,350 \text{ cal} = 105 \text{ cal} (x \text{ g})$$

$$\text{Wt } H_2O = 156 \text{ g}$$

12. What x-ray wavelength would be diffracted at an angle of 21.16° if the interplanar spacing is 2.50 Å?

Soln: The Bragg equation is required:

$$\lambda = 2d \sin \theta$$

$$= 2(2.50 \text{ Å}) \sin 21.16 = (5.0 \text{ Å}) (0.3610)$$

$$= 1.80 \text{ Å}$$

17. At what angle would 0.711-Å x rays be diffracted from planes spaced at 1.77 Å?

Soln: The Bragg equation is used to find θ:

$$\lambda = 2d \sin \theta$$

$$0.711 \text{ Å} = 2(1.77 \text{ Å}) \sin \theta$$

$$\sin \theta = 0.2008$$

$$\theta = 11.6°$$

UNIT VI

Solutions

13

INTRODUCTION

This chapter treats the general physical-chemical properties of solutions and ways in which solutions are used in chemical reactions. Units used for expressing solution concentrations are treated in detail.

    The problems emphasize these units of concentration for preparing solutions, for calculating freezing-point depressions and boiling-point elevations, and for determining quantities of reagents required for reactions.

FORMULAS AND DEFINITIONS

Concentrated vs dilute A concentrated solution contains a higher ratio of solute to solvent than a dilute solution does.

Gram-equivalent weight (GEW) The amount of substance that donates or accepts one mole (6.022×10^{23}) of electrons or protons, H^+, or neutralizes one mole of negative or positive charges. Specifically, for

Acids and bases: The GEW of an acid or base used in a reaction is the GFW of the substance divided by the number of protons or hydroxide ions transferred per molecule of acid or base, respectively:

$$GEW = \frac{GFW \text{ of acid or base}}{\text{no. of } H^+ \text{ or } OH^- \text{ transferred}}$$

Oxidation-reduction processes: The GEW of a substance is its GFW divided by the number of electrons transferred per formula unit of the substance:

$$GEW = \frac{\text{GFW of substance}}{\text{no. of electrons transferred}}$$

Related to reactions: Components of a chemical reaction are generally related in the following manner:

$$\text{No. GEW Reactant}_1 = \text{no. GEW Reactant}_2 = \cdots = \text{no. GEW Reactant}_n$$

$$= \text{no. GEW Product}_1 = \cdots = \text{no. GEW Product}_n$$

The following are units of concentration.

Molality (m)   The molality of a solution is the number of moles of solute dissolved per kilogram of solvent used:

$$m = \frac{\text{no. moles of solute}}{\text{no. kg of solvent}}$$

Molarity (M)   The molarity of a solution is the number of moles of solute dissolved per liter of solution:

$$M = \frac{\text{no. moles of solute}}{\text{vol. of solution, in liters}}$$

Normality (N)   The normality of a solution is the number of gram-equivalent weights of solute dissolved per liter of solution:

$$N = \frac{\text{no. gram-equivalent weights solute}}{\text{vol. of solution, in liters}}$$

Percentage   The weight percentage of a component in solution is the weight of the component divided by the total weight of solution and the fraction multiplied by 100 to get percentage:

$$\% = \frac{\text{wt of component}}{\text{wt of solution}} \times 100$$

Mole fraction (X)   The mole fraction of a component in solution is the number of moles of a component divided by the total number of moles of all components in solution:

$$X = \frac{\text{no. moles of component}}{\text{total moles of all components}}$$

Dilution formulas   A volume of concentrated solution can be diluted. A larger volume of solution thereby is formed, in which the number of moles or gram-equivalent weights of solute has remained constant; only the volume of solution has changed. Therefore, the following two relations exist, which are useful for preparing solution.

Molarity $\qquad (M_{conc})(V_{conc}) = (M_{dil})(V_{dil})$

Normality $\qquad (N_{conc})(V_{conc}) = (N_{dil})(V_{dil})$

PROBLEMS

Note: Unless otherwise specified, the word solution *refers to a system in which water is used as the solvent for a specified solute.*

1. How many grams of a 6.00% NaCl solution by weight are necessary to yield 80.0 g of NaCl?

   Soln: The percentage of a component in solution is the weight of the component divided by the total weight of the solution and the quotient multiplied by 100 for percentage. Since 6.00% of the solution weight equals 80.0 g,

   $$(0.0600) \text{ (Wt NaCl solution)} = 80.0 \text{ g}$$

   $$\text{Wt NaCl solution} = \frac{80.0 \text{ g}}{0.0600} = 1330 \text{ g}$$

2. How many grams of $HC_2 H_3 O_2$ are contained in 60.0 ml of acetic acid of specific gravity 1.065 and containing 58.0% $HC_2 H_3 O_2$ by weight?

   Soln: The specific gravity factor, which also is included in many of the following problems, is used to define the weight of a specified volume of solution. The specific gravity of this solution is 1.065, which means that this solution weighs 1.065 times as much as an equal volume of water. Also, it is assumed that the density of water is 1.00 g/ml, or 1000 g/liter. Therefore, the density of this solution is 1.065 times 1.0 g/ml, or times 1000 g/liter.

   $$\text{Wt of 60.0 ml of acetic acid} = 60.0 \text{ ml} \times 1.065 \text{ g/ml}$$

   $$= 63.9 \text{ g}$$

   $$\text{Wt } HC_2 H_3 O_2 = (0.58) (63.9 \text{ g}) = 37.1 \text{ g}$$

3. Calculate the weight of solute in each of the following solutions: (a) 2.40 liters of 0.650 M $HClO_4$; (b) 125 ml of 0.0250 M $C_{12} H_{22} O_{11}$.

   Soln: Molarity M is defined as the number of moles of solute per liter of solution and can be expressed

   $$M = \frac{\text{no. moles of solute}}{\text{volume of solution (liters)}}$$

   where no. moles = wt solute/GFW. Then

   $$M = \frac{\text{wt solute/GFW}}{V \text{ (liters)}}$$

   $$\text{Wt solute} = (M) (V_1) (\text{GFW})$$

   (a) GFW of $HClO_4$ = 100.5 g

   $$\text{Wt of } HClO_4 = (0.650 \, M) (2.40 \text{ liters}) (100.5 \text{ g}) = 157 \text{ g}$$

(b) GFW of $C_{12}H_{22}O_{11}$ = 342 g

Wt of $C_{12}H_{22}O_{11}$ = (0.0250 M) (0.125 liter) (342 g) = 1.07 g

4. How much sulfuric acid (98.0% by weight) is needed in the preparation of 250 g of 25.0% solution of the acid by weight?

Soln: The concentrated acid solution is 98.0 percent H_2SO_4 and 2.0 percent water by weight. This problem, and others, involves a percentage purity factor that must be respected in solving the problem. Therefore the weight of H_2SO_4 needed is

Wt of H_2SO_4 = (0.250) (250 g) = 62.5 g

The weight of the 98.0 percent H_2SO_4 solution required is

(0.980) (Wt of H_2SO_4 solution) = 62.5 g

$$Wt = \frac{62.5 \text{ g}}{0.980} = 63.8 \text{ g}$$

The solution would be prepared by slowly adding 63.8 g of the H_2SO_4 solution to 186.2 g of water.

6. Calculate the molarity of each of the following solutions:
(a) 15.6 g of CsOH in 1.50 liters of aqueous solution.
(d) 7.0 millimoles of I_2 in 100 ml of carbon tetrachloride solution.

Soln: (a) GFW of CsOH = 149.9 g

$$M \text{ (CsOH)} = \frac{\text{no. moles}}{V \text{ (liters)}}$$
$$= \frac{15.6 \text{ g} \times (1.0 \text{ mole}/149.9 \text{ g})}{1.50 \text{ liters}} = 0.0694 \text{ } M$$

(d) The ratio of moles to liters is equivalent to the ratio of millimoles to milliliters.

$$M \text{ (I}_2\text{)} = \frac{7.0 \text{ m-moles}}{100 \text{ ml}} = 0.070 \text{ } M$$

8. How would you prepare 90.0 g of a 1.50 percent $Ba(OH)_2$ solution by weight starting with $Ba(OH)_2 \cdot 8 H_2O$ and water?

Soln: Many substances contain water of hydration, for example, $Ba(OH)_2 \cdot 8 H_2O$. The water of hydration can be treated as a percentage purity relation, as in Problem 4. First, calculate the percentage of $Ba(OH)_2$ in the hydrate:

$$\% \text{ Ba(OH)}_2 = \frac{\text{GFW Ba(OH)}_2}{\text{GFW of hydrate}} \times 100 = \frac{171.3}{315.3} \times 100 = 54.3\%$$

Wt $Ba(OH)_2$ needed = (0.015) (90.00 g) = 1.35 g

$$Wt \text{ Ba(OH)}_2 \cdot 8 H_2O = \frac{1.35 \text{ g}}{0.543} = 2.49 \text{ g}$$

To prepare 90.00 g of solution, add 2.49 g of the hydrate to 87.51 g of water.

10. How many liters of HBr measured at S.T.P. are required in the preparation of 5.00 liters of 0.500 M hydrobromic acid. *Note:* See the solution to Problem 15 for help if necessary.

11. Calculate the molarity of a 500-ml volume of solution containing 3.0 g of 95.0 % sulfuric acid, $H_2 SO_4$.

Soln: $$M = \frac{\text{No. moles}}{V \text{ (liters)}} = \frac{\text{wt solute/GFW}}{V \text{ (liters)}}$$

Wt $H_2 SO_4$ = (0.95) (3.0 g) = 2.85 g;   GFW $H_2 SO_4$ = 98.0 g

$$M = \frac{2.85 \text{ g} \times (1 \text{ mole/98.0 g})}{0.50 \text{ liter}} = 0.058 \ M$$

12. Calculate the molarity of a solution of α-fructose, $C_6 H_{12} O_6$, for which the density is 1.06 g/ml and which contains 15.00% of the compound by weight.

Soln: First, assume a solution volume of 1.00 liter. The weight of the solution is

Wt = 1.06 g/ml $\times$ 1000 ml/liter = 1060 g/liter

Then

Wt of $C_6 H_{12} O_6$ = (0.15) (1060 g/liter) = 159 g/liter

GFW of $C_6 H_{12} O_6$ = 180.0 g

$$M = \frac{159 \text{ g} \times (1 \text{ mole/180.0 g})}{1.00 \text{ liter}} = 0.883 \ M$$

15. The specific gravity of a 20.0% by weight solution of aqueous ammonia is 0.925. What volume of ammonia gas (S.T.P.) would be required in the preparation of 2.50 liters of this solution?

Soln: The solution in this problem is prepared, as in Problem 10, by dissolving a gas in water. The weight of gas required is calculated by using the molar volume relation 22.4 liters gas = 1.0 mole measured at S.T.P.

Wt NH_3/liter = (0.200) (925 g/liter) = 185.0 g/liter; GFW NH_3 =

17.03 g; no. moles NH_3/liter = 185.0 g/liter $\times \dfrac{1 \text{ mole}}{17.03 \text{ g}}$ = 10.86

V_{NH_3} at S.T.P. = 10.86 moles/liter $\times$ 2.50 liters $\times \dfrac{22.4 \text{ liters}}{\text{mole}}$

= 608 liters

16. It is desired to produce 1.00 liter of 0.400 M sulfuric acid by diluting 12.00 M sulfuric acid. The concentrated acid is poured cautiously into the water. Calculate the volume of the concentrated acid and the volume of water required in the dilution.

Soln: A concentrated solution can be used to prepare a less concentrated solution by adding water to a predetermined volume of the concentrated solution. In principle, the number of moles of solute remains unchanged in the dilution process. Therefore no. moles (concentrated solution used) = no. moles (dilute solution), or

$$(M_1)(V_1) = (M_2)(V_2)$$

$$(12.00\ M)(V) = (0.400\ M)(1.000\ \text{liter})$$

$$V = \frac{(0.400\ M)(1.000\ \text{liter})}{12.00\ M}$$

$$= 0.0333\ \text{liter} = 33.3\ \text{ml acid}$$

Add 33.3 ml of H_2SO_4; 967 ml of H_2O.

18. Concentrated hydrochloric acid is 37.0% HCl by weight and has a specific gravity of 1.19. What volume of this acid must be diluted to 500 ml to produce a 0.300 M solution of the acid?

Soln: First, calculate the weight of concentrated HCl per milliliter of the solution:

$$\text{Wt of HCl/ml} = (0.37)(1.19\ \text{g/ml}) = 0.440\ \text{g/ml}$$

Second, calculate the weight of HCl required to prepare the solution:

GFW of HCl = 36.45 g

$$\text{Wt of HCl} = (0.300\ M)(0.500\ \text{liter})(36.45\ \text{g/mole}) = 5.47\ \text{g}$$

The volume of concentrated solution required is

$$V = 5.47\ \text{g} \times \frac{1.00\ \text{ml}}{0.44\ \text{g}} = 12.4\ \text{ml}$$

21. A 10.3-g sample of $K_2CO_3 \cdot 1.5\ H_2O$ is dissolved in 150 g of water. Calculate the percentage by weight of K_2CO_3 in the solution and the molality of the solution in terms of K_2CO_3.

Soln: Calculate the weight of K_2CO_3 in 10.3 g of $K_2CO_3 \cdot 1.5\ H_2O$.

GFW of K_2CO_3 = 138.2 g;   GFW of $K_2CO_3 \cdot 1.5\ H_2O$ = 165.2 g

$$\% K_2CO_3 \text{ in } K_2CO_3 \cdot 1.5\ H_2O = \frac{138.2}{165.2} \times 100 = 83.65\%$$

43

Wt of K_2CO_3 = (0.8365) (10.3 g) = 8.62 g

% solution K_2CO_3 = $\dfrac{8.62 \text{ g}}{160.3 \text{ g}}$ × 100 = 5.37%

24. A standard solution of NaOH has a molarity of 0.09987. What is the molarity of a nitric acid solution if 10.0 ml of it neutralizes 24.6 ml of the NaOH solution?

Soln: The neutralization reaction is $HNO_3 + NaOH \rightarrow NaNO_3 + H_2O$. For complete reaction, the number of moles of HNO_3 must equal the number of moles of NaOH used:

$$(M_{NaOH})(V_{NaOH}) = (M_{HNO_3})(V_{HNO_3})$$

$$M(HNO_3) = \frac{(0.09987)(24.6 \text{ ml})}{10.0 \text{ ml}} = 0.246 \text{ } M$$

27. Calculate the gram-equivalent weight of each of the reactants in the following:
(a) $H_2SO_4 + 2LiOH \rightarrow Li_2SO_4 + 2H_2O$
(b) $KOH + KHCO_3 \rightarrow K_2CO_3 + H_2O$
(d) $Mg + 2HCl \rightarrow MgCl_2 + H_2$

Soln: (a) $\underline{H_2SO_4}$. Each molecule transfers two H^+.

$$\text{GEW of } H_2SO_4 = \frac{GFW}{2} = \frac{98.0 \text{ g}}{2} = 49.0 \text{ g}$$

$\underline{LiOH}$. Each molecule transfers one OH^-.

$$\text{GEW of LiOH} = GFW = 23.9 \text{ g}$$

(b) $\underline{KOH}$ and $\underline{KHCO_3}$. Each molecule transfers one OH^- and one H^+, respectively.

$$\text{GEW of KOH} = GFW \text{ of KOH} = 56.1 \text{ g}$$

$$\text{GEW of } KHCO_3 = GFW \text{ of } KHCO_3 = 100.1 \text{ g}$$

(d) $\underline{Mg}$. One atom of magnesium loses two electrons.

$$\text{GEW of Mg} = \frac{GFW}{2} = 12.2 \text{ g}$$

$\underline{HCl}$. One molecule of HCl furnishes one H^+.

$$\text{GEW of HCl} = GFW = 36.5 \text{ g}$$

28. Calculate the normality of each of the following solutions:
(a) 4.0 gram-equivalent weights of LiI in 2.0 liters of solution.
(b) 2.5 milligram-equivalent weights of $MgCl_2$ in 50 ml of solution.

Soln: Normality is calculated from its definition as follows: GEW/liter = m-GEW/ml.

(a) N (LiI) $= \dfrac{4.0 \text{ GEW}}{2.0 \text{ liters}} = 2.0 \, N$

(b) N (MgCl$_2$) $= \dfrac{2.5 \text{ m-GEW}}{50 \text{ ml}} = 0.050 \, N$

30. The chloride in 50.0 ml of dilute hydrochloric acid was precipitated using an excess of silver nitrate. The weight of AgCl formed was 0.682 g. Calculate the normality of the hydrochloric acid solution.

Soln: The number of gram-equivalent weights of HCl equals the number of gram-equivalent weights of AgCl produced by the reaction.

$$\text{AgNO}_3 + \text{HCl} \rightarrow \underline{\text{AgCl}} + \text{HNO}_3$$

GEW of AgCl = GFW = 143.33 g

No. GEW of AgCl = $(V_{\text{HCl}})(N_{\text{HCl}})$

$$0.682 \text{ g AgCl} \times \frac{1 \text{ GEW}}{143.33 \text{ g}} = (0.050 \text{ liter})(N)$$

$N_{\text{HCl}} = 0.0952$

32. What volume of 0.400 M HCl would be required to react completely with 2.60 g of sodium carbonate? Na$_2$CO$_3$ + 2HCl $\rightarrow$ 2NaCl + CO$_2$ + H$_2$O.

Soln: In this reaction, one molecule of Na$_2$CO$_3$ neutralizes two molecules of HCl. The equivalent weights are

GEW of HCl = GFW = 36.5 g;   GEW of Na$_2$CO$_3$ $= \dfrac{\text{GFW}}{2} = 53.0$ g

No. GEW of Na$_2$CO$_3$ = no. GEW of HCl

$$2.60 \text{ g} \times \frac{1 \text{ GEW}}{53.0 \text{ g}} = (0.40 \, N)(V_{\text{HCl}})$$

0.123 liter $= V_{\text{HCl}} = 123$ ml

34. A 75.0-ml volume of gaseous ammonia measured at 25.0° and 753 mm pressure was absorbed in 50.0 ml of water. (a) What is the normality of the ammonia solution? (b) How many ml of 0.100 N sulfuric acid would be required in the neutralization of this aqueous ammonia? Assume no change in volume when the gaseous ammonia is added to the water.

Soln: (a) The GEW of NH$_3$ is determined on the basis of the neutralization of NH$_3$ with acid. One molecule of NH$_3$ can accept one proton from a molecule of acid: NH$_3$ + H$^+$ $\rightarrow$ NH$_4$$^+$. On this basis, the GEW of NH$_3$ equals its formula weight. Therefore, the number of GEW quantities of NH$_3$ equals the number of moles of NH$_3$, and the normality of the solution equals its molarity. Calculate the molarity, and hence the normality of the NH$_3$ solution.

$$V \text{ of } NH_3 \text{ at S.T.P.} = 75.0 \text{ ml} \times \frac{753}{760} \times \frac{273}{298} = 68.1 \text{ ml}$$

$$\text{No. moles } NH_3 = 68.1 \text{ ml} \times \frac{1.0 \text{ liter}}{1000 \text{ ml}} \times \frac{1 \text{ mole}}{22.4 \text{ liters}} = 0.00304 \text{ mole}$$

No. GEW = 0.00304

$$N(NH_3) = \frac{0.00304 \text{ GEW}}{0.0500 \text{ liter}} = 0.0608 \, N$$

(b) The concept of gram-equivalent weight relates all components of reactions to each other: the number of gram-equivalent weights for each component in a reaction equals the numbers for all other components. In this case,

$$\text{No. GEW of } NH_3 = \text{no. GEW of } H_2SO_4$$

or

$$(N_{NH_3})(V_{NH_3}) = (N_{H_2SO_4})(V_{H_2SO_4})$$
$$0.00304 \text{ GEW} = (0.100 \, N)(V)$$
$$V = 0.0304 \text{ liter}$$
$$V = 30.4 \text{ ml}$$

36. What volume of 0.300 M H_2O_2 would be required to oxidize 0.634 g Na_2SO_3 of 98.0% purity? ($Na_2SO_3 + H_2O_2 \rightarrow NaSO_4 + H_2O$)

Soln: This problem is most easily solved by using molar relations, since one mole of H_2O_2 reacts with one mole of Na_2SO_3.

Wt of Na_2SO_3 = (0.98)(0.634 g) = 0.621 g

GFW of Na_2SO_3 = 126.1 g

No. moles of Na_2SO_3 = no. moles of H_2O_2

$$0.621 \text{ g} \times \frac{1 \text{ mole}}{126.1 \text{ g}} = (0.300 \, M)(V \text{ liters})$$

$$0.0164 \text{ liter} = V_{H_2O_2} = 16.4 \text{ ml}$$

39. A 0.4238-gram sample of an acid was dissolved in water. It took 34.7 ml of 0.100 M NaOH solution to neutralize the acid. What is the gram-equivalent weight of the acid?

Soln: No. GEW of Acid = no. GEW of NaOH

No. GEW of NaOH = (0.0347 liter)(0.100 N) = 0.00347

Then

$$0.00347 \text{ GEW acid} = 0.4238 \text{ g}$$

$$\text{GEW of acid} = \frac{0.4238}{0.00347} = 122 \text{ g}$$

40. A mixture of NaBr and KI weighing 0.93 g was dissolved in water and then the halides were precipitated as silver salts, weighing together 1.64 g. Calculate the composition of the original mixture.

Soln: This problem involves two unknowns, the weight of NaBr and the weight of KI. The reactions involved are

$$\text{NaBr} + \text{AgNO}_3 \rightarrow \underline{\text{AgBr}} + \text{NaNO}_3$$

$$\text{KI} + \text{AgNO}_3 \rightarrow \underline{\text{AgI}} + \text{KNO}_3$$

According to the data, two relations exist:

$$\text{Wt of NaBr} + \text{Wt KI} = 0.93 \text{ g}$$

$$\text{Wt of AgBr} + \text{Wt AgI} = 1.64 \text{ g}$$

GFW: NaBr = 102.9; AgBr = 187.7 g; KI = 166.0 g; AgI = 234.8 g. Let x = Wt NaBr and $0.93 - x$ = Wt KI. The weights of AgBr and AgI produced are

$$\text{Wt of AgBr} = x \text{ g NaBr} \times \frac{1 \text{ mole}}{102.9 \text{ g}} \times \frac{1 \text{ mole AgBr}}{1 \text{ mole NaBr}} \times \frac{187.7 \text{ g}}{\text{mole}}$$

$$= \frac{(187.7)(x \text{ g})}{102.9}$$

$$\text{Wt of AgI} = (0.93 - x) \text{ g KI} \times \frac{1 \text{ mole}}{166.0 \text{ g}} \times \frac{1 \text{ mole AgI}}{1 \text{ mole KI}} \times \frac{234.8 \text{ g}}{\text{mole}}$$

$$= \frac{218.4 - 234.8x}{166} \text{ g}$$

Adding the two equations yields

$$\frac{187.7x}{102.9} + \frac{218.4 \text{ g} - 234.8x}{166} = 1.64 \text{ g}$$

Multiplying both sides of the equation by (102.9) (166) gives

$$31{,}158x + 22{,}473.4 \text{ g} - 24{,}160x = 28{,}013.5 \text{ g}$$

$$6998x = 5540 \text{ g}$$

$$x = 0.79 \text{ g} = \text{Wt of NaBr}$$

$$(0.93 - 0.79) \text{ g} = \text{Wt KI} = 0.14 \text{ g}$$

43. How would you prepare a 4.21 molal aqueous solution of propyleneglycol ($C_3H_8O_2$)? What would be the freezing point of this solution?

Soln: A 4.21 molal solution contans 4.21 moles of solute per kilogram of solvent.

GFW of $C_3H_8O_2$ = 76.0 g

Wt of $C_3H_8O_2$ = 76.0 g/mole $\times$ 4.21 moles = 320 g

Dissolve 320 g of propyleneglycol in 1000 g of water.

The change in freezing point of the solvent is

$$\Delta f\, p = (4.21\ m)\,(1.86°/m) = 7.83°$$

Freezing point = $-7.83°$C.

44. If 35.7 g of the nonelectrolyte $C_6H_3Cl_3$ is dissolved in 220 g of chloroform, what is (a) the freezing point of the solution and (b) the boiling point of the solution at 760 mm? For chloroform, the freezing point is $-63.5°$, the boiling point is $61.26°$, K_f is $4.68°$, and K_b is $3.63°$.

Soln: Calculate the number of moles and then the fp and bp:

GFW of $C_6H_3Cl_3$ = 181.4 g

$$\text{No. moles} = 35.7\ g \times \frac{1\ \text{mole}}{181.4\ g} = 0.197\ \text{mole}$$

(b) $$\Delta bp = m\,K_b = (\frac{0.197\ \text{mole}}{0.220\ \text{kg}})\,(3.63°/m) = 3.25°$$

bp = $61.26° + 3.25° = 64.5°$C

45. A 4.305-gram sample of a nonelectrolyte is dissolved in 105 g of water. The solution freezes at $-1.23°$C. Calculate the molecular weight of the substance.

Soln: The freezing point of a solution is related to the molality of the solution and the freezing-point depression constant (K_f) of the solvent: $\Delta fp = m\,K_f$.

$$m = \frac{\text{no. moles solute}}{\text{no. kg solvent}} = \frac{\text{wt solute/GFW}}{\text{no. kg solvent}}$$

$$\Delta fp = \frac{\text{wt solute/GFW}}{\text{no. kg solvent}} \times K_f$$

Solve for GFW:

$$GFW = \frac{(\text{wt solute})\,(K_f)}{(\Delta fp)\,(\text{no. kg solvent})}$$

$$= \frac{(4.305\ g)\,(1.86°/m)}{(1.23°)\,(0.105\ kg)}$$

$$= 62.0\ g$$

48. How many grams of formaldehyde, CH_2O, must be added to a liter of water to prevent freezing at $5.00°$F?

Soln: The freezing point of the solution is 5.00°F, or −15.0°C.

GFW of CH_2O = 30.0 g;   1.0 liter H_2O = 1.0 kg

$$\Delta fp = \left(\frac{\text{wt solute}/\text{GFW}}{\text{no. kg}}\right) K_f$$

$$\text{Wt of } CH_2O = \frac{(\Delta fp)\,(\text{no. kg})\,(\text{GFW})}{K_f}$$

$$= \frac{(15°)\,(1.0\text{ kg})\,(30.0\text{ g})}{1.86°/m} = 242\text{ g}$$

51. Calculate the mole fraction of solute and solvent for a 2.1 molal aqueous solution.

Soln: In this particular system, 2.1 moles of solute are dissolved per kilogram of water. The mole fractions are

$$X_{\text{solute}} = \frac{\text{moles solute}}{\text{moles solute + moles solvent}}$$

$$\text{No. moles } H_2O = 1000\text{ g} \times \frac{1\text{ mole}}{18.0\text{ g}} = 55.56$$

$$X_{\text{solute}} = \frac{2.1\text{ moles}}{(2.1 + 55.6)\text{ moles}} = 0.036$$

$$X_{H_2O} = \frac{55.6\text{ moles}}{57.7\text{ moles}} = 0.96$$

55. Concentrated hydrochloric acid is 37.0% HCl by weight and has a specific gravity of 1.19. Calculate (a) the molarity, (b) the molality, (c) the normality, and (d) the mole fraction of HCl and of H_2O.

Soln: A liter of this solution weighs 1190 g.

Wt of HCl/liter = (0.37) (1190 g/liter) = 440.3 g

GFW of HCl = 36.5 g

(a) M = no. moles HCl/liter

$$= \frac{440.3\text{ g} \times (1\text{ mole}/36.5\text{ g})}{1.0\text{ liter}} = 12.1\text{ molar}$$

(b) $m = \dfrac{\text{no. moles HCl}}{\text{no. kg } H_2O}$

$$= \frac{12.1\text{ moles}}{(1190\text{ g} - 440.3\text{ g})/1000}$$

$$= \frac{12.1\text{ moles}}{0.75\text{ kg}} = 16.1\text{ molal}$$

(c) GFW of HCl = GEW and $M_{HCl} = N_{HCl}$

(d) No. moles water = $750 \text{ g} \times \dfrac{1 \text{ mole}}{18.0 \text{ g}} = 41.67$

$$X_{HCl} = \dfrac{12.1 \text{ moles}}{(12.1 + 41.7) \text{ moles}} = \dfrac{12.1 \text{ moles}}{53.8 \text{ moles}} = 0.225 \text{ mole fraction HCl}$$

$$X_{H_2O} = \dfrac{41.7 \text{ moles}}{53.8 \text{ moles}} = 0.775 \text{ mole fraction } H_2O$$

UNIT VII

Solutions of Electrolytes; Colloids

14

INTRODUCTION

This chapter treats the nature of solutions in terms of colligative properties, the solubility of substances in solvents, and colloid behavior.

FORMULAS AND DEFINITIONS

Weak electrolyte   Substance whose aqueous solutions are poor conductors of electricity. Also, a substance in which only a small fraction of its particles are dissociated in aqueous solution.

Strong electrolyte   Substance whose aqueous solutions are good conductors of electricity. Also, a substance in which effectively all the solute particles are dissociated into ions in aqueous solution.

Activity coefficient   Henry's law and Raoult's law define the behavior of an ideal solution. However, real solutions deviate from ideal behavior and the extent of deviation is defined in terms of an activity coefficient, f, as follows:

Activity of solute = activity coefficient $\times$ actual molality of solution

$$\alpha = fm$$

Activity of solute in solution   The activity of a solute in a solution is a measure of its effective molality in the solution, that is, the real effect of a solute in solution compared with the ideal effect.

PROBLEMS

1. If 200 ml of 0.200 M NaCl, 300 ml of 0.100 M Na$_2$SO$_4$, 100 ml of 0.500 M K$_2$SO$_4$ and 200 ml of 0.300 M NaNO$_3$ are mixed, what is the molarity of each ion in the resulting solution? Assume complete dissociation of the substances.

   Soln: These substances are strong electrolytes and to good approximation we can assume complete dissociation of each substance in solution. We can calculate the number of moles of each substance, and subsequently the number of moles of each ion in solution.

   NaCl. (200 ml) (0.200 M) = 40.0 m-moles

   $$NaCl \rightarrow Na^+ + Cl^-$$

   No. moles Na$^+$ = No. moles Cl$^-$= 40.0 m-moles

   Na$_2$SO$_4$. (300 ml) (0.100 M) = 30.0 m-moles

   $$Na_2SO_4 \rightarrow 2Na^+ + SO_4^{2-}$$

   No. moles Na$^+$ = 60.0 m-moles; no. moles SO$_4^{2-}$= 30.0 m-moles.

   K$_2$SO$_4$. (100 ml) (0.500 M) = 50.0 m-moles.

   $$K_2SO_4 \rightarrow 2K^+ + SO_4^{2-}$$

   No. moles K$^+$ = 100.0 m-moles; no. moles SO$_4^{2-}$= 50.0 m-moles.

   NaNO$_3$. (200 ml) (0.300 M) = 60.0 m-moles

   $$NaNO_3 \rightarrow Na^+ + NO_3^-$$

   No. moles Na$^+$ = no. moles NO$_3^-$= 60.0 m-moles.

   $$\text{Molarity Na}^+ = \frac{160 \text{ m-moles}}{800 \text{ ml}} = 0.200 \ M \text{ in Na}^+$$

   $$\text{Molarity Cl}^- = \frac{40.0 \text{ m-moles}}{800 \text{ ml}} = 0.0500 \ M \text{ in Cl}^-$$

   $$\text{Molarity SO}_4^{2-} = \frac{80.0 \text{ m-moles}}{800 \text{ ml}} = 0.100 \ M \text{ in SO}_4^{2-}$$

   $$\text{Molarity K}^+ = \frac{100.0 \text{ m-moles}}{800 \text{ ml}} = 0.125 \ M \text{ in K}^+$$

   $$\text{Molarity NO}_3^- = \frac{60.0 \text{ m-moles}}{800 \text{ ml}} = 0.0750 \ M \text{ in NO}_3^-$$

2. What would be the approximate freezing point of a 0.42 molal aqueous solution of cesium iodide?

Soln: The approximate freezing point of the solution can be calculated by assuming that the activity coefficient equals 1.0, that is, that the compound completely dissociates in solution and there is no particle interaction.

$$\Delta fp = (K_f)\,(m)\,(\text{no. ions per formula unit})$$

$$= (1.86°/m)\,(0.42)\,(2) = 1.6°C$$

Freezing point $= -1.6°C$.

3. A 0.500-molal aqueous solution of an acid HA shows a freezing point of $-1.04°C$. What is the percent dissociation of this acid?

Soln: The formula HA is used to represent a weak acid substance that dissociates, as follows: $HA + H_2O \rightarrow H_3O^+ + A^-$. Only a small percentage of weak acid molecules are dissociated in aqueous solution. The freezing-point depression depends on the total number of particles in solution. Therefore, the effective molality of the solution arises from the total number of moles of particles present or the activities of each ion present:

Total moles = moles HA + moles H^+ + moles A^-

$$\Delta fp = K_f\, m_{\text{effective}}$$

$$m_{\text{effective}} = \frac{1.04°}{1.86} = 0.559 \text{ molal}$$

Let x = no. moles of HA that dissociate

$$= \text{no. moles of } H^+ \text{ and } A^- \text{ produced}$$

$$(0.500 - x) + x + x = 0.559$$

$$x = 0.059 \text{ mole}$$

$$\% \text{ dissociation} = \frac{0.059 \text{ mole}}{0.500} \times 100 = 11.8\%$$

4. A solution is made by dissolving 0.498 g of potassium bromide in 100.0 g of water. The solution is observed to freeze at $-0.150°C$. If one assumes that the activity coefficient of the potassium ion is equal to that of the bromide ion, what is the activity coefficient of the potassium ion under these conditions?

Soln: The activity coefficient, f, is calculated from the effective molality and the molality based on the amount of KBr added.

$$\text{Molality KBr} = \frac{0.498 \text{ g} \times (1 \text{ mole}/119 \text{ g})}{0.100 \text{ kg } H_2O} = 0.0418\, m$$

$$\Delta fp = K_f m$$

$$m_{\text{effective}} = \frac{0.150°}{1.86} = 0.0806\, m$$

53

If all the KBr dissociates, the effective molality will be twice the actual molality, or

$$2(0.0418) = 0.0836 \; m$$

$$\text{activity} = f(\text{molality})$$

$$f = \frac{0.0806}{0.0836} = 0.964$$

6. The activity coefficient for the ions in 0.100 molal HI is 0.818 at 25.0°C. Calculate the activities of these ions.

Soln: Activity $= f(\text{molality})$

Activity H^+ and $I^- = (0.818)(0.100 \; m) = 0.0818$ molal

UNIT VIII

Oxidation-Reduction Reactions

16

INTRODUCTION

This chapter deals exclusively with problems involving oxidation-reduction reactions. The balancing of redox reactions is developed by the change in oxidation number method and by the ion-electron method. These methods are applied to the gram-equivalent weights of oxidizing and reducing agents and to determining the normality of these solutions.

FORMULAS AND DEFINITIONS

Oxidation   A loss of electrons; can be defined as an increase in the positive number of an atom or ion.

Reduction   A gain of electrons; can be defined as a decrease in the positive oxidation number of atom or ion.

Oxidizing agent (or oxidant)   A substance that increases the oxidation number of elements.

Reducing agent (reductant)   A substance that decreases the oxidation number of elements.

Gram-equivalent weights of oxidizing and reducing agents   The amount of substance that donates or accepts one mole of electrons (6.023×10^{23}).

Stoichiometric calculations involving redox reactions Identical in principle with those reactions of acids and bases. The number of milliequivalents of oxidizing agent used must equal the number of milliequivalents of reducing agent used.

The following two operations relate to balancing redox equations.

Change in oxidation number method Based on the concept that in a redox reaction the total increase in units of positive oxidation number must equal the total decrease in units of negative oxidation number. Consider the following reaction as an example:

$$MnO_4^- + C_2O_4^{2-} + H^+ \rightarrow Mn^{2+} + CO_2 + H_2O$$

Assign oxidation numbers to each element, determine those elements that undergo oxidation or reduction, and then write half-reactions that indicate the oxidation number changes. In this case, the half-reactions are

$$C_2^{3+} \rightarrow 2C^{4+} + 2e^- \qquad \text{(gain in positive oxidation number, 2)}$$

$$Mn^{7+} \rightarrow Mn^{2+} - 5e^- \qquad \text{(decrease in positive oxidation number, 5)}$$

To balance the equation, we must select the proper number of manganese-containing and carbon-containing ions so that the total increase of electrons will equal the total decrease. This can be done by choosing the smallest factor common to both 2 and 5 and multiplying each half-reaction by the multiple of each in the factor. For these half-reactions, 10 is the common factor and the half-reactions must be multiplied by 5 and 2, respectively.

$$5[C_2^{3+} \rightarrow 2C^{4+} + 2e^-] \qquad 5 \times 2e^- = 10e^-$$

$$2[Mn^{7+} \rightarrow Mn^{2+} - 5e^-] \qquad 2 \times 5e^- = 10e^-$$

The proper coefficients for those substances changing oxidation state become:

$$2MnO_4^- + 5C_2O_4^{2-} + H^+ \rightarrow Mn^{2+} + 10CO_2 + H_2O$$

The remaining factors in the equation can be balanced by inspection, yielding

$$2MnO_4^- + 5C_2O_4^{2-} + 16H^+ \rightarrow 2Mn^{2+} + 10CO_2 + 8H_2O$$

Ion-electron method Based on combining half-reactions, one half representing the oxidation step and the other the reduction step. This method is very useful in electrochemical cells, for which generally the half-reactions are known. The same problem used in the change in oxidation number method will be used to illustrate this technique.

The oxalate ion is oxidized to carbon dioxide:

$$C_2O_4^{2-} \rightarrow 2CO_2 + 2e^- \tag{1}$$

The equation is balanced by inspection. The oxidizing agent in this reaction, permanganate, is reduced in acid to manganous ion and water:

$$MnO_4^- + 8H^+ + 5e^- \to Mn^{2+} + 4H_2O \qquad (2)$$

Again the balancing is done by inspection.

It is now obvious that multiplying Equation (1) by 5 and Equation (2) by 2 will make the number of electrons lost by oxalate the same as the number gained by permanganate. The two equations are added to give the following:

$$
\begin{array}{lll}
5C_2O_4^{2-} \to 10CO_2 & & + 10e^- \\
2MnO_4^- + 16H^+ \quad + 10e^- \to 2Mn^{2+} & & + 8H_2O \\
\hline
2MnO_4^- + 5C_2O_4^{2-} + 16H^+ \to 2Mn^{2+} + 10CO_2 & + 8H_2O
\end{array}
$$

Both methods give the same result, as we should expect.

QUESTIONS

4. Balance the following redox equations by the "change of oxidation number" and/or the "ion-electron" method:

Soln: (a) $H_2SO_4 + HBr \to SO_2\uparrow + Br_2\uparrow + H_2O$

$$
\begin{array}{lll}
S^{6+} & \to \ S^{4+} & - 2e^- \\
2Br^- & \to \ Br_2^{\,0} & + 2e^- \\
\hline
& S^{4+} & + Br_2^{\,0}
\end{array}
$$

$H_2SO_4 + 2HBr \to SO_2\uparrow + Br_2\uparrow + 2H_2O$

(c) $MnO_4^- + S^{2-} + H_2O \to MnO_2 + S + OH^-$

$$
\begin{array}{lll}
2[Mn^{7+} \to & Mn^{4+} & - 3e^-] \\
3[S^{2-} \to & S^0 & + 2e^-] \\
\hline
& 2Mn^{4+} & + 3S^0
\end{array}
$$

$2MnO_4^- + 3S^{2-} + 4H_2O \to 2MnO_2 + 3S + 8OH^-$

(e) $Cu + H^+ + NO_3^- \to Cu^{2+} + NO_2\uparrow + H_2O$

$$
\begin{array}{lll}
Cu & \to \ Cu^{2+} & + 2e^- \\
2[N^{5+} \to & N^{4+} & - 1e^-] \\
\hline
& Cu^{2+} & + 2N^{4+}
\end{array}
$$

$Cu + 4H^+ + 2NO_3^- \to Cu^{2+} + 2NO_2\uparrow + 2H_2O$

(g) $Cu + H^+ + NO_3^- \rightarrow Cu^{2+} + NO\uparrow + H_2O$

$$3[Cu^0 \rightarrow Cu^{2+} + 2e^-]$$
$$\underline{2[N^{5+} \rightarrow N^{2+} - 3e^-]}$$
$$3Cu^{2+} + 2N^{2+}$$

$$3Cu + 8H^+ + 2NO_3^- \rightarrow 3Cu^{2+} + 2NO\uparrow + 4H_2O$$

(i) $H_2SO_4 + HI \rightarrow H_2S + I_2 + H_2O$

$$4[2I^- \rightarrow I_2{}^0 + 2e^-]$$
$$\underline{S^{6+} \rightarrow S^{2-} - 8e^-}$$
$$4I_2 + S^{2-}$$

$$H_2SO_4 + 8HI \rightarrow H_2S + 4I_2 + 4H_2O$$

(k) $OH^- + Cl_2 \rightarrow ClO_3^- + Cl^- + H_2O$

$$Cl^0 \rightarrow Cl^{5+} + 5e^-$$
$$\underline{5[Cl^0 \rightarrow Cl^{1-} - 1e^-]}$$
$$Cl^{5+} + 5Cl^-$$

$$6OH^- + 3Cl_2 \rightarrow ClO_3^- + 5Cl^- + 3H_2O$$

(m) $MnO_4{}^{2-} + H_2O \rightarrow MnO_4^- + OH^- + MnO_2$

$$2[Mn^{6+} \rightarrow Mn^{7+} + 1e^-]$$
$$\underline{Mn^{6+} \rightarrow Mn^{4+} - 2e^-}$$
$$2Mn^{7+} + Mn^{4+}$$

$$3MnO_4{}^{2-} + 2H_2O \rightarrow 2MnO_4^- + 4OH^- + \underline{MnO_2}$$

Here, balancing charges is convenient: "6$^-$" on the left side indicates that four ions (OH^-) must be added to the right side to balance the charge.

5. Balance the following redox equations by the "change in oxidation number" and/or the "ion-electron" method:

Soln: (a) $MnO_4{}^{2-} + Cl_2 \rightarrow MnO_4^- + Cl^-$

$$2[Mn^{6+} \rightarrow Mn^{7+} + 1e^-]$$
$$\underline{Cl_2{}^0 \rightarrow 2Cl^- - 2e^-}$$
$$2Mn^{7+} + 2Cl^-$$

$$2MnO_4{}^{2-} + Cl_2 \rightarrow 2MnO_4^- + 2Cl^-$$

(c) $HBrO \rightarrow H^+ + Br^- + O_2\uparrow$

$$2[Br^{1+} \rightarrow Br^{1-} \quad - 2e^-]$$
$$\underline{2[O^{2-} \rightarrow O_2^0 \quad + 4e^-]}$$
$$2Br^- \quad + O_2$$

$2HBrO \rightarrow 2H^+ + 2Br^- + O_2\uparrow$

(e) $CuS + H^+ + NO_3^- \rightarrow Cu^{2+} + \underline{S} + NO\uparrow + H_2O$

$$3[S^{2-} \rightarrow S^0 \quad + 2e^-]$$
$$\underline{2[N^{5+} \rightarrow N^{2+} \quad - 3e^-]}$$
$$3S^0 \quad + 2N^{2+}$$

$3CuS + 8H^+ + 2NO_3^- \rightarrow 3Cu^{2+} + 3\underline{S} + 2NO\uparrow + 4H_2O$

(g) $C + HNO_3 \rightarrow NO_2\uparrow + H_2O + CO_2\uparrow$

$$C^0 \quad \rightarrow C^{4+} \quad + 4e^-$$
$$\underline{4[N^{5+} \rightarrow N^{4+} \quad - 1e^-]}$$
$$C^{4+} \quad + 4N^{4+}$$

$C + 4HNO_3 \rightarrow 4NO_2\uparrow + 2H_2O + CO_2\uparrow$

(i) $NO_3^- + Zn + OH^- + H_2O \rightarrow NH_3\uparrow + Zn(OH)_4^{2-}$

$$4[Zn^0 \rightarrow Zn^{2+} \quad + 2e^-]$$
$$\underline{N^{5+} \quad \rightarrow N^{3-} \quad - 8e^-}$$
$$4Zn^{2+} + N^{3-}$$

$NO_3^- + 4Zn + 7OH^- + 6H_2O \rightarrow NH_3\uparrow + 4Zn(OH)_4^{2-}$

After balancing the half-reaction, it is convenient to count ionic charges to decide the number of ions OH^- and H_2O molecules required. The right side of the equation contains $4(2^-)$, or 8^-, charges; therefore, with the 1^- charge on NO_3^-, seven additional negative charges must be added by way of seven OH^-. Then, only the amount of water must be considered.

(k) $H_2S + H_2O_2 \rightarrow \underline{S} + H_2O$

$$S^{2-} \quad \rightarrow S^0 \quad + 2e^-$$
$$\underline{O_2^{2-} \quad \rightarrow 2O^{2-} \quad - 2e^-}$$
$$S^0 \quad + 2O^{2-}$$

$H_2S + H_2O_2 \rightarrow \underline{S} + 2H_2O$

(m) $Al + H^+ \rightarrow Al^{3+} + H_2 \uparrow$

$$2[Al^0 \rightarrow Al^{3+} + 3e^-]$$
$$\underline{3[2H^+ \rightarrow H_2^0 - 2e^-]}$$
$$2Al^{3+} + 3H_2^0$$

$$2Al + 6H^+ \rightarrow 2Al^{3+} + 3H_2 \uparrow$$

6. Complete and balance the following equations. When the reaction occurs in acidic solution, H^+ and/or H_2O may be added on either side of the equation, as necessary, to balance the equation properly; when the reaction occurs in basic solution, OH^- and/or H_2O may be added, as necessary, on either side of the equation. No indication of the acidity of the solution is given if neither H^+ nor OH^- is involved as a reactant or product.

Soln: (a) $Ag + NO_3^- \rightarrow Ag^+ + NO$ (acidic solution)

$$3[Ag^0 \rightarrow Ag^+ + 1e^-]$$
$$\underline{N^{5+} \rightarrow N^{2+} - 3e^-}$$
$$3Ag^+ + N^{2+}$$

$$4H^+ + 3Ag + NO_3^- \rightarrow 3Ag^+ + NO + 2H_2O$$

(c) $H_2S + I_2 \rightarrow S + I^-$

$$S^{2-} \rightarrow S^0 + 2e^-$$
$$\underline{I_2^0 \rightarrow 2I^- - 2e^-}$$
$$S^0 + 2I^-$$

$$H_2S + I_2 \rightarrow S + 2I^- + 2H^+$$

(e) $Ag^+ + AsH_3 \rightarrow Ag + H_3AsO_3$ (acidic solution)

$$As^{3-} \rightarrow As^{3+} + 6e^-$$
$$\underline{6[Ag^+ \rightarrow Ag^0 - 1e^-]}$$
$$As^{3+} + 6Ag^0$$

Remember charge balance! It is convenient to balance ionic charge along with oxidation number charge in all examples in Questions 6 and 7.

$$3H_2O + 6Ag^+ + AsH_3 \rightarrow 6Ag + H_3AsO_3 + 6H^+$$

(g) $CN^- + MnO_4^- \rightarrow CNO^- + MnO_2$ (basic solution)

$$3[N^{3-} \rightarrow N^- + 2e^-]$$
$$\underline{2[Mn^{7+} \rightarrow Mn^{4+} - 3e^-]}$$
$$3N^- + 2Mn^{4+}$$

$$H_2O + 3CN^- + 2MnO_4^- \rightarrow 3CNO^- + 2MnO_2 + 2OH^-$$

7. Complete and balance the following equations (see instructions for Question 6):

Soln: (a) $Fe^{2+} + MnO_4^- \rightarrow Fe^{3+} + Mn^{2+}$ (acidic solution)

$$5[Fe^{2+} \rightarrow Fe^{3+} + 1e^-]$$

$$\underline{Mn^{7+} \rightarrow Mn^{2+} - 5e^-}$$

$$5Fe^{3+} + Mn^{2+}$$

$$5Fe^{2+} + MnO_4^- \rightarrow 5Fe^{3+} + Mn^{2+}$$

It is convenient to balance ionic charge along with oxidation number charge. Left side, 9^+; right side, 17^+. Add eight H^+ to the left side to balance the charge, and four H_2O to the right side for atomic balance.

$$5Fe^{2+} + MnO_4^- + 8H^+ \rightarrow 5Fe^{3+} + Mn^{2+} + 4H_2O$$

(c) $Hg_2Cl_2 + NH_3 \rightarrow Hg + HgNH_2Cl + NH_4^+ + Cl^-$

$$Hg^+ \rightarrow Hg^{2+} + 1e^-$$

$$\underline{Hg^+ \rightarrow Hg^0 - 1e^-}$$

$$Hg^{2+} + Hg^0$$

$$Hg_2Cl_2 + 2NH_3 \rightarrow Hg + HgNH_2Cl + NH_4^+ + Cl^-$$

(e) $MnO_2 + Cl^- \rightarrow Mn^{2+} + Cl_2$ (acidic solution)

$$2Cl^- \rightarrow Cl_2{}^0 + 2e^-$$

$$\underline{Mn^{4+} \rightarrow Mn^{2+} + 2e^-}$$

$$Cl_2{}^0 + Mn^{2+}$$

$$MnO_2 + 2Cl^- + 4H^+ \rightarrow Mn^{2+} + Cl_2 + 2H_2O$$

(g) $Fe^{2+} + Cr_2O_7{}^{2-} \rightarrow Fe^{3+} + Cr^{3+}$ (acidic solution)

$$6[Fe^{2+} \rightarrow Fe^{3+} + 1e^-]$$

$$\underline{Cr_2{}^{6+} \rightarrow 2Cr^{3+} - 6e^-]}$$

$$6Fe^{3+} + 2Cr^{3+}$$

$$6Fe^{2+} + Cr_2O_7{}^{2-} + 14H^+ \rightarrow 6Fe^{3+} + 2Cr^{3+} + 7H_2O$$

(i) $Sn^{2+} + HgCl_2 + Cl^- \rightarrow SnCl_6{}^{2-} + Hg_2Cl_2$

$$Sn^{2+} \rightarrow Sn^{4+} + 2e^-$$

$$\underline{2Hg^{2+} \rightarrow Hg_2{}^{2+} - 2e^-}$$

$$Sn^{4+} + Hg_2{}^{2+}$$

$$Sn^{2+} + 2HgCl_2 + 4Cl^- \rightarrow SnCl_6{}^{2-} + \underline{Hg_2Cl_2}$$

(k) $CH_2O + [Ag(NH_3)_2]^+ \rightarrow Ag + HCO_2^- + NH_3$ (basic solution)

$$2[Ag^{1+} \rightarrow Ag^0 + 1e^-]$$

$$\underline{C^0 \quad \rightarrow \quad C^{2+} \quad + 2e^-}$$

$$2Ag^0 \quad + C^{2+}$$

$$3OH^- + CH_2O + 2[Ag(NH_3)_2]^+ \rightarrow 2Ag + HCO_2^- + 4NH_3 + 2H_2O$$

8. Complete and balance the following equation (see instructions for Question 6). (This is an exceptionally difficult equation to balance.)

$$Pb(N_3)_2 + Cr(MnO_4)_2 \rightarrow$$

$$Cr_2O_3 + MnO_2 + Pb_3O_4 + NO \quad \text{(basic solution)}$$

Soln: $15Pb(N_3)_2 + 44Cr(MnO_4)_2 \rightarrow$

$$22Cr_2O_3 + 88MnO_2 + 5Pb_3O_4 + 90NO$$

PROBLEMS

1. What weight of aluminum would be necessary to reduce all the cadmium in 240 ml of a 0.150 M solution of $CdCl_2$? $3Cd^{2+} + 2Al \rightarrow 3Cd + 2Al^{3+}$.

  Soln: No. m-GEW oxidizing agent = no. m-GEW reducing agent
Cadmium is gaining two electrons and is the oxidizing agent causing the loss of three electrons from aluminum:

    No. GEW of Al = 2(0.150 M) (0.240 liter) = 0.072

    Wt of Al = (0.072 GEW) (9 g/GEW) = 0.648 g

3. Calculate the equivalent weight of the oxidizing and reducing agents in reactions (d), (i), (j), (l), and (n) of Question 4.

  Soln: For Question 4 (i),

$$H_2SO_4 + 8HI \rightarrow H_2S + 4I_2 + 4H_2O$$

Oxidizing agent H_2SO_4: each molecule transfers 8 e^-; therefore the equivalent weight equals the formula weight divided by 8.

$$GEW = \frac{98 \text{ g}}{8} = 12.3 \text{ g}$$

Reducing agent HI: each molecule transfers one e^-; therefore, the equivalent weight is the atomic weight, 127.9 g.

    All others are calculated in a similar manner; that is, calculate the formula weight for each agent, then divide by the number of electrons transferred for that particular agent.

6. A solution of $(NH_4)_2SO_4 \cdot FeSO_4 \cdot 6H_2O$ containing 1.05 grams of the compound is titrated with an acidic $Na_2Cr_2O_7$ solution. If 41.6 milliliters of the $Na_2Cr_2O_7$ solution is needed, what is the normality of the $Na_2Cr_2O_7$ solution? $Fe^{2+} + Cr_2O_7^{2-} + H^+ \rightarrow Cr^{3+} + Fe^{5+} + H_2O$ (unbalanced).

Soln: The iron is oxidized from the 2^+ oxidation state to 3^+.

No. moles of $(NH_4)_2SO_4 \cdot FeSO_4 \cdot 6H_2O$ = No. GEW

$$1.05 \text{ g} \times \frac{1 \text{ GEW}}{392.1 \text{ g}} = 0.00268 \text{ GEW}$$

GEW of $(NH_4)_2SO_4 \cdot FeSO_4 \cdot 6H_2O$ = liters $\times$ normality

$$\text{Normality } Na_2Cr_2O_7 = \frac{\text{GEW of } (NH_4)_2SO_4 \cdot FeSO_4 \cdot 6H_2O}{\text{liters } Na_2Ca_2O_7}$$

$$N = \frac{0.00268}{0.0416} = 0.0644$$

8. The chlorine content of a solution of chlorine water is to be determined. A 50.0-ml sample (sp. gr. = 1.02) of this chlorine water is treated with an excess of potassium iodide. To titrate the liberated iodine, 34.5 ml of $0.100 \; N \; Na_2S_2O_3$ is required. Calculate the percentage of chlorine by weight in the chlorine water. (Chlorine water is a solution of free chlorine in water.) $Cl_2 + 2I^- \rightarrow 2Cl^- + I_2$.

Soln: No. m-GEW of $Na_2S_2O_3$ = No. m-GEW of I_2 or Cl_2

No. GEW of Cl_2 = (0.0345 liter) (0.100 N) = 0.00345

Since each Cl_2 molecule present gains 2e⁻,

$$\text{GEW of } Cl_2 = \frac{70.9 \text{ g}}{2} = 35.45 \text{ g}$$

Amount Cl_2 present = No. equivalents $\times$ GEW

$$= (0.00345)(35.45) = 0.1223 \text{ g}$$

The weight of the solution is 50 ml $\times$ 1.02 g/ml = 51.0 g. The percentage of Cl_2 is

$$\frac{0.1223 \text{ g}}{51.0 \text{ g}} \times 100 = 0.240\%$$

9. State for each of the following whether the initial substance is an oxidizing agent or a reducing agent and make the requested conversion.
   (a) 0.3 mole of H_2S to equivalent weights of H_2S (product is S).
   (c) 1.6 milliequivalent weights of HNO_3 to millimoles of HNO_3 (product is NH_3).

Soln: (a) $S^{2-} \rightarrow S^0 + 2e^-$ (reducing agent)

$$\text{GEW of } H_2S = \frac{GFW}{2} \text{ or } \frac{\text{no. moles}}{2}$$

$$\text{No. GEW} = 0.3 \text{ mole} \times 2 \text{ GEW/mole} = 0.6$$

(c) $N^{5+} \rightarrow N^{3-} - 8e^-$ (oxidizing agent)

1 GFW of HNO_3 = 8 GEW of HNO_3

$$\text{No. m-moles} = 1.6 \text{ mew} \times \frac{1 \text{ m-mole}}{8 \text{ mew}} = 0.20 \text{ millimole}$$

UNIT IX

Chemical Equilibrium
17

INTRODUCTION

Both the theory and the applications of chemical equilibrium are discussed in this chapter. The broad concepts involved are treated, using the collision theory of reaction. The reaction mechanisms and the effect of temperature, catalysis, and concentration on the rate of reaction are covered in detail so that the law of chemical equilibrium can be developed. The equilibrium constant is introduced and shown to be independent of changing concentrations of reactants and products, and of catalysts. The theory of heterogeneous and homogeneous equilibria has important considerations for modern industry and also as a basis for further research.

DEFINITIONS AND FORMULAS

Equilibrium   A condition in a system when two opposing reactions occur simultaneously at the same rate.

Rate of chemical reaction   The number of moles of a substance that disappear or are formed by a reaction in a unit of time. In a simple reaction in which the stoichiometric coefficients are all one for the products and reactants, the rate R is directly proportional to the concentrations of the reacting substances.

For the reaction $A + B \rightleftharpoons C + D$, the relation between the reaction rate and the molar concentrations (indicated by brackets) can be written in the form of a rate equation, $R_1 = k_1 [A] [B]$. In this expression, k_1 is the forward rate constant for the formation of product. The opposing reaction is the formation of reactants; its rate is $R_2 = k_2 [C] [D]$. At equilibrium, the opposing rates are equal. Setting the rates equal gives

$$k_1 [A] [B] = k_2 [C] [D] \quad \text{or} \quad \frac{k_1}{k_2} = \frac{[C] [D]}{[A] [B]}$$

Since the ratio of the two constants is a constant, the basis is laid for writing a new constant, $K_e = k_1/k_2$, where K_e is the equilibrium constant. For the general chemical reaction, $mA + nB + \cdots + xC + yD \ldots$, the equilibrium constant can be written

$$K_e = \frac{[C]^x [D]^y}{[A]^m [B]^n}$$

Determination of equilibrium constants In determining concentrations at equilibrium, it sometimes happens that there is obtained a quadratic equation that must be solved for x (some unknown concentration). To use the quadratic formula, the quadratic equation must be arranged in the form $ax^2 + bx + c = 0$. Direct substitution of the values of a, b, and c into the quadratic formula,

$$x = \frac{-b \pm \sqrt{b^2 - 4ac}}{2a}$$

followed by solution will give two values of x. Generally, one of these values will not make chemical sense in the framework of the problem and must be discarded, leaving one value that is reasonable.

PROBLEMS

1. Nitrogen reacts with hydrogen to give ammonia according to the equation $N_2 + 3H_2 \rightleftharpoons 2NH_3$. An equilibrium mixture of the above substances at $400°C$ was found to contain 0.45 mole of nitrogen, 0.63 mole of hydrogen, and 0.24 mole of ammonia per liter. Calculate the equilibrium constant for the system.

   Soln: This is a simple problem requiring the use of the equilibrium constant for the reaction $N_2 + 3H_2 \rightleftharpoons 2NH_3$. Set up the mass action expression and substitute the concentrations 0.45 mole/liter N_2, 0.63 mole/liter H_2, and 0.24 mole/liter NH_3.

$$K_e = \frac{[NH_3]^2}{[N_2] [H_2]^3} = \frac{[0.24]^2}{[0.45] [0.63]^3} = \frac{0.0576}{(0.45) (0.250)} = 0.51$$

3. The equilibrium constant for the gaseous reaction $H_2 + I_2 \rightleftharpoons 2HI$ is 50.2 at $448°C$. Calculate the number of grams of HI that are in equilibrium with 1.25 moles of H_2 and 63.5 grams of iodine at this temperature.

Soln: The number of grams of HI can be calculated from the number of moles of HI determined from the equilibrium constant expression.

$$K_e = \frac{[HI]^2}{[H_2][I_2]} = 50.2$$

$$[HI]^2 = K_e[H_2][I_2]$$

$$[HI]^2 = \frac{(50.2)(1.25 \text{ moles})(63.5 \text{ g})}{253.8 \text{ g/mole } I_2} = 15.70 \text{ moles}^2$$

$$[HI] = (15.70)^{1/2} = 3.962 \text{ moles}$$

$$\text{Wt of HI} = 3.962 \text{ moles} \times 127.9 \text{ g/mole HI} = 507 \text{ g}$$

4. The equilibrium constant for the reaction $CO + H_2O \rightleftharpoons CO_2 + H_2$ is 5.0 at a given temperature.

(a) On analysis, an equilibrium mixture of the above substances at the given temperature was found to contain 0.20 mole of CO, 0.30 mole of water vapor, and 0.90 mole of H_2 in a liter. How many moles of CO_2 were there in the equilibrium mixture?

(b) Maintaining the same temperature, additional H_2 was added to the system, and some water vapor removed by drying. A new equilibrium mixture was thereby established which contained 0.40 mole of CO, 0.30 mole of water vapor, and 1.2 mole of H_2 in a liter. How many moles of CO_2 were in the new equilibrium mixture? Compare the value with the quantity in part (a) and discuss whether the second value is reasonable. Explain how it is possible for the water vapor concentration to be the same in the two equilibrium solutions even though some was removed before the second equilibrium was established.

Soln: Use is again made of the equilibrium constant expression to solve for an unknown concentration of a 1-liter system:

(a) $$K_e = 5.0 = \frac{[CO_2][H_2]}{[CO][H_2O]}$$

$$[CO_2] = \frac{K_e[CO][H_2O]}{[H_2]} = \frac{5.0\,(0.20)\,(0.30)}{(0.90)} = 0.33$$

(b) $$[CO_2] = \frac{K_e[CO][H_2O]}{[H_2]}$$

$$= \frac{5.0\,(0.40)\,(0.30)}{1.2} = 0.50$$

Although the numbers calculated for the amount of CO_2 present in part (a) and part (b) are correct, they are not consistent when the two parts as a whole are compared. The total number of moles of carbon-containing material (the carbon content should not

67

change in amount) is 0.33 + 0.20 = 0.53 mole in part (a) and 0.50 + 0.40 = 0.90 mole in part (b). However, it is not inconsistent to find the same amount of water present in the two parts since the equilibrium could shift to form more water once some has been removed, as in part (b).

5. At about 990°C, K_e for the reaction $H_2(g) + CO_2(g) \rightleftharpoons H_2O(g) + CO(g)$ is 1.6. Calculate the number of moles of each component in the final equilibrium system obtained from adding 1.00 mole of H_2 and 1.00 mole of CO_2 to a 5.00-liter reactor at 990°C.

Soln: $H_2 + CO_2 \rightleftharpoons H_2O + CO$; the initial concentrations are $[H_2] = 1.00$ mole/5.00 liter = 0.20; $[CO_2]$ = 1.00 mole/5.00 liter.

$$K_e = 1.6 = \frac{[H_2O]\,[CO]}{[H_2]\,[CO_2]}$$

Since $K_e > 1$, the reaction must proceed to the right; thus the concentration of H_2O and CO must increase. Let x = number of moles of H_2 and CO_2 that react to form H_2O and CO. The concentrations are

t	H_2	CO_2	H_2O	CO
0	0.20	0.20	0	0
eq	$0.20 - x$	$0.20 - x$	x	x

$$K_e = \frac{(x)\,(x)}{(0.20 - x)\,(0.20 - x)} = 1.6$$

$$x^2 = 1.6(0.20 - x)\,(0.20 - x)$$

$$= 0.064 - 0.64x + 1.6x^2$$

$$0.6x^2 - 0.64x + 0.064 = 0$$

This equation is of the form $ax^2 + bx + c = 0$ and can be solved for x by substitution into the quadratic formula,

$$x = \frac{-b \pm \sqrt{b^2 - 4ac}}{2a}$$

$$= \frac{+0.64 \pm \sqrt{0.410 - 0.154}}{1.2}$$

$$= \frac{0.64 \pm 0.506}{1.2} = 0.96 \text{ or } 0.112$$

These are the amounts found in one liter. In 5 liters, we should have 0.96 mole/liter × 5 liter = 4.8 moles or 0.112 mole/liter × 5 liter = 0.56 mole. Since 4.8 moles is greater than the initial concentration, it cannot be a correct solution for x. Therefore, 0.56 mole is the amount of H_2O and CO formed and 1.00 − 0.56 = 0.44 mole of H_2 and CO_2 remaining.

6. Ethanol and acetic acid interact to form ethyl acetate and water, according to the equation

$$C_2H_5OH + CH_3COOH \rightleftharpoons CH_3COOC_2H_5 + H_2O$$

When one mole each of ethanol (C_2H_5OH) and acetic acid (CH_3COOH) are allowed to react at $100°C$ in a sealed tube, equilibrium is established when one-third of a mole of each of the reactants remains. Calculate K_e.

Soln: $C_2H_5OH + HC_2H_3O_2 \rightleftharpoons CH_3COOC_2H_5 + H_2O$. To see more clearly the relation of initial and equilibrium concentrations, you will find the block form convenient.

t	C_2H_5OH	HOAc	EtOAc	H_2O
0	1.0	1.0	0	0
eq	$\frac{1}{3}$	$\frac{1}{3}$	$\frac{2}{3}$	$\frac{2}{3}$

$$K_e = \frac{[CH_3COOC_2H_5][H_2O]}{[C_2H_5OH][HC_2H_3O_2]} = \frac{(\frac{2}{3})(\frac{2}{3})}{(\frac{1}{3})(\frac{1}{3})} = 4$$

8. If the rate of a reaction doubles for every ten-degree rise in temperature, how much faster would the reaction proceed at $55°$ than at $25°$? at $100°$ than at $25°$?

Soln: The rate doubles for each $10°C$ rise in temperature. Fifty-five degrees is a $30°$ increase over $25°$. Thus the rate doubles three times, or 2^3 (rate at $25°$) = 8 times faster. One hundred degrees is a $75°$ increase over $25°$. Thus the rate doubles 7.5 times, or $2^{7.5}$ (rate at $25°$) = 180 times faster. This procedure gives a rough approximation!

12. An equilibrium mixture of N_2, H_2, and NH_3 in a 1.00-liter vessel is found to contain 0.300 mole of N_2, 0.400 mole of H_2, and 0.100 mole of NH_3. How many moles of H_2 must be introduced into the vessel in order to double the equilibrium concentration of NH_3 if the temperature remains unchanged?

Soln: This problem gives enough information to find the equilibrium constant. Then, through the equilibrium condition, the amount of nitrogen under the new equilibrium conditions is found, leaving only the amount of hydrogen present as an unknown.

$$N_2 + 3H_2 \rightleftharpoons 2NH_3$$

$$K_e = \frac{[NH_3]^2}{[N_2][H_2]^3} = \frac{(0.100)^2}{(0.300)(0.400)^3} = \frac{0.010}{(0.300)(0.064)} = 0.521$$

Since $[NH_3]$ must be doubled, an additional 0.100 mole of NH_3 must be formed, the N of NH_3 coming from the N_2.

1 mole $N_2 \rightarrow$ 2 moles NH_3; 0.050 mole $N_2 \rightarrow$ 0.100 mole NH_3

The new conditions are

$$[N_2] = 0.300 - 0.050 = 0.250 \text{ mole}$$

$$[H_2]^3 = \frac{[NH_3]^2}{K[N_2]} = \frac{(0.200)^2}{(0.521)(0.250)} = 0.307$$

$$[H_2] = (0.307)^{1/3} = 0.675$$

An additional 0.15 mole of H_2 is required to produce 0.10 mole of NH_3. The initial amount of H_2 is 0.400 mole minus 0.150 mole, or 0.250 mole. Therefore, the amount of H_2 to be added is $(0.675 - 0.250)$ mole, or 0.425 mole.

13. At 25°C and atmospheric pressure, the partial pressures in an equilibrium mixture of N_2O_4 and NO_2 are $N_2O_4 = 0.70$ atmosphere; and $NO_2 = 0.30$ atmosphere. Calculate the partial pressures of these two gases when they are in equilibrium at 9.0 atmospheres and 25°C.

Soln: First calculate the equilibrium constant in terms of pressure, then proceed as previously.

$$p_{N_2O_4} = 0.70 \text{ atm}; p_{NO_2} = 0.3 \text{ atm}. \quad N_2O_4 \rightleftharpoons 2NO_2$$

$$K = \frac{(p_{NO_2})^2}{p_{N_2O_4}} = \frac{(0.3)^2}{0.7} = 0.13$$

For 9.0 atm, let $x = p_{NO_2}$ and $9.0 - x = p_{N_2O_4}$.

$$K = 0.13 = \frac{(x)^2}{9.0 - x} \quad \text{and} \quad x^2 = 1.17 - 0.13x$$

$$x^2 + 0.13x - 1.17 = 0$$

Substitution into the quadratic formula gives

$$x = \frac{-0.13 \pm \sqrt{0.017 + 4.68}}{2}$$

$$= \frac{-0.13 \pm 2.17}{2}$$

$$= \frac{2.04}{2} = 1.02$$

The second answer obtained, -1.1, has no chemical meaning; hence the pressure of NO_2 is 1.0 atm; the pressure of N_2O_4 is $(9.0 - 1.0)$ atm = 8.0 atm.

14. Consider the gas phase reaction $N_2O_4(g) \rightleftharpoons 2NO_2(g)$. At a certain temperature, K_e for this reaction is 1.1×10^{-5}. If 0.20 mole of N_2O_4 is dissolved in 400 ml of chloroform and the above reaction allowed to come to equilibrium, (a) what will be the NO_2 concentration, and (b) what will be the percentage of dissociation of the original N_2O_4?

Soln: $N_2O_4(g) \rightleftharpoons 2NO_2(g);$ $\quad K_e = 1.1 \times 10^{-5} = \dfrac{[NO_2]^2}{[N_2O_4]}$

Since $K < 1$, the reaction will proceed to the side of the reactants. Thus, some NO_2 must form, starting from pure N_2O_4. Let $x =$ the amount dissociated and $[0.20/0.400 - x] =$ concentration of N_2O_4. Thus,

$$1.1 \times 10^{-5} = \frac{x^2}{0.5 - x}$$

Here, x in the denominator is small enough in comparison to 0.5 to be ignored.

$x = 2.3 \times 10^{-3}$ mole/liter = conc. of NO_2

(b) The percentage of dissociation is the amount dissociated times 100. Thus, $2.3 \times 10^{-3} \times 100 = 0.23\%$.

18. In a 3.0-liter vessel, the following equilibrium partial pressures are measured: N_2, 190 mm; H_2, 317 mm; NH_3, 1000 mm. Hydrogen is removed from the vessel until the partial pressure of nitrogen, at equilibrium, is equal to 250 mm. Calculate the partial pressures of the other substances under the new conditions.

Soln: $N_2 + 3H_2 \rightleftharpoons 2NH_3$

$$K_e = \frac{(p_{NH_3})^2}{(p_{N_2})(p_{H_2})^3} = \frac{(1000)^2}{(190)(317)^3} = 1.65 \times 10^{-4}$$

$p_{N_2} = 250$ mm increased from 190 mm, a change of 60 mm. Since $N_2 \rightarrow 2NH_3$, p_{NH_3} should decrease by 2(60 mm), or 120 mm. $p_{NH_3} = 1000 - 120 = 880$ mm. After rearrangement,

$$(p_{H_2})^3 = \frac{(p_{NH_3})^2}{(p_{N_2})K_e}$$

$$= \frac{(880)^2}{(250)(1.65 \times 10^{-4})}$$

$$= 1.88 \times 10^7$$

$$p_{H_2} = 266 \text{ mm}$$

19. What is the minimum weight of $CaCO_3$ which is required to establish equilibrium at a certain temperature in a 6.50-liter container if the equilibrium constant is 0.050 mole/liter for the decomposition reaction of $CaCO_3$ at that temperature? Justify the units given in the problem for the equilibrium constant. The equation for the reaction is $CaCO_3(s) \rightleftharpoons CaO(s) + CO_2(g)$.

Soln: This problem involves condensed phases in contact with a gas. The equilibrium expression can be written

$$K_e = \frac{[CO_2] \, [CaO]}{[CaCO_3]} = 0.050$$

Since CaO and $CaCO_3$ are solids, their concentrations are unchanging so long as the slightest amount of solid is present. Hence, the expression reduces to $K = [CO_2] = 0.050$ mole/liter. Since there are 6.5 liters, the total number of moles present is 0.050 mole/liter $\times$ 6.5 liter = 0.33 mole.

Wt of $CaCO_3$ = 100 g/mole $\times$ 0.33 mole = 33 g

22. At high temperatures, the reaction of hydrogen and oxygen to form water is reversible and achieves equilibrium. Under such conditions, a one-liter container holding pure hydrogen at a pressure of 0.80 atmosphere and another one-liter container holding pure oxygen at a pressure of 20.0 atmospheres were connected and opened to each other so that the gases could mix and come to chemical equilibrium with water vapor. Then, the partial pressure of oxygen was measured and found to be *essentially* 10.0 atmospheres. (Assume that the quantity of oxygen which reacts is negligible compared to the quantity originally present.) The equilibrium constant for the reaction at the temperature of the experiment has a value of 2.50 atm^{-1}. (a) Write the equilibrium constant expression for the reaction. (b) What are the equilibrium partial pressures of hydrogen and water? (c) Check whether the assumption made regarding the quantity of oxygen which reacts compared to the quantity originally present is justifiable. Could a similar assumption be justified for hydrogen? Explain.

Soln: (a) $2H_2 + O_2 \rightleftharpoons 2H_2O$; $\quad K_e = \dfrac{[H_2O]^2}{[H_2]^2 \, [O_2]}$

(b) Since the partial pressure of oxygen is essentially 10 atm and its original pressure (taking into account the doubling of volume) is 10.0 atm, the sum of p_{H_2} and p_{H_2O} must equal 0.40 atm (half of 0.80 atm). This must be true since $2H_2 \rightarrow 2H_2O$, leaving the pressure unchanged. Let $x = p_{H_2}$; $0.4 - x = p_{H_2O}$.

$$K = \frac{(0.4 - x)^2}{(x^2) \, (10)} = 2.5$$

From the quadratic formula, the two solutions are $x = 0.07$, and $x = -4.8/48$; the latter must be ruled out. Therefore, $p_{H_2} = 0.07$ atm; $p_{H_2O} = 0.40 - 0.07 = 0.33$ atm.

23. For the reaction $A \rightarrow B + C$ the following data were obtained at $30°C$:

Experiment	[A] (in mole/liter)	Rate (in mole $\cdot$ liter^{-1} $\cdot$ hr^{-1}
1	0.170	0.0500
2	0.340	0.100
3	0.680	0.200

(a) What is the rate equation and order of the reaction?
(b) Calculate k for the reaction.

Soln: (a) It is intended that the data for each separate experiment given in this problem be substituted in the general rate equation to determine how m, the order of the reaction, varies. For the reaction $A \rightarrow B + C$, the rate equation is $R = k[A]^m$. Then, for

$$\text{Exp. 1, } 0.0500 = k[0.170]^m$$

$$\text{Exp. 2, } 0.100 = k[0.340]^m$$

$$\text{Exp. 3, } 0.200 = k[0.680]^m$$

Since k and m refer to the same reaction, they must retain the same values throughout the three experiments. In going from Exp. 1 to Exp. 2, a doubling of the concentration doubles the rate. The same is true for Exp. 2 and Exp. 3. The rate equation holds for these data only if m equals 1. Therefore, the rate is first-order.

(b) Using the data from Exp. 1, we have

$$R = k[A]^1 = 0.0500 = k[0.170]$$

$$k = \frac{0.0500}{0.170} = 0.294 \text{ hr}^{-1}$$

25. The reaction of compound A to give compounds C and D was found to be second-order in A and second-order overall. The rate constant for the process was determined to be 2.42 liter $\cdot$ mole^{-1} $\cdot$ sec^{-1}. If the initial concentration is 0.0500 mole/liter, what is the value of $t_{1/2}$?

Soln: For a second-order reaction in which there is only one reactant, the governing equation is

$$\frac{1}{[A]} - \frac{1}{[A_0]} = kt$$

For a half-life determination, we substitute $t = t_{1/2}$ and $[A] = \frac{1}{2}[A_0]$. This reduces to

$$t_{1/2} = \frac{1}{k[A_0]}$$

$$= \frac{1}{(2.42 \text{ liter} \cdot \text{mole}^{-1} \cdot \text{sec}^{-1})[0.0500 \text{ liter} \cdot \text{mole}^{-1}]}$$

$$= 8.26 \text{ sec}$$

27. The half-life of a reaction involving compound A was found to be 8.50 min when the initial concentration of A was 0.150 mole/liter. How long will it take for the concentration to drop to 0.0300 mole/liter if the reaction is (a) first-order overall and first-order with respect to A? (b) second-order overall and second-order with respect to A? In your calculations, indicate the units for each quantity, and show how the units cancel to give the proper units in the final answer.

Soln: (a) To determine the time required for the concentration to drop to a particular value for a first-order reaction, we need to know the rate constant. For our problem $t_{1/2} = 0.693/k = 8.50$ min.

$$k = \frac{0.693}{8.50 \text{ min}} = 0.0815 \text{ min}^{-1}$$

Substituting this and the initial concentration in the basic equation governing first-order reactions

$$\log \frac{[A_0]}{[A]} = \frac{kt}{2.303}$$

yields

$$\log \frac{0.150 \text{ mole/liter}}{0.0300 \text{ mole/liter}} = \frac{(0.0815 \text{ min}^{-1})t}{2.303}$$

$$t = \frac{0.699 \, (2.303)}{0.0815 \text{ min}^{-1}} = 19.7 \text{ min}$$

(b) For a second-order equation, the governing equation is

$$\frac{1}{[A]} - \frac{1}{[A_0]} = kt$$

Here k must again be calculated using the data for the problem:

$$\frac{1}{0.150/2 \text{ mole/liter}} - \frac{1}{0.150 \text{ mole/liter}} = k \cdot 8.50 \text{ min}$$

$$k = \frac{13.33 \text{ moles/liter} - 6.66 \text{ moles/liter}}{8.50 \text{ min}}$$

$$= \frac{6.66 \text{ moles/liter}}{8.50 \text{ min}}$$

$$= 0.784 \text{ mole} \cdot \text{liter}^{-1} \cdot \text{min}^{-1}$$

Substituting k and the new concentration gives:

$$\frac{1}{0.0300} - \frac{1}{0.150} = (0.784 \text{ mole} \cdot \text{liter}^{-1} \cdot \text{min}^{-1})t$$

$$33.33 - 6.66 = 26.67 = 0.784\, t$$

$$t = \frac{26.67}{0.784 \text{ min}^{-1}} = 34.0 \text{ min}$$

UNIT X

Ionic Equilibria of Weak Electrolytes

18

INTRODUCTION

Weak electrolytes are substances that form aqueous solutions containing ions derived from the electrolyte and also undissociated electrolyte molecules. In most cases, only a very small fraction of the total number of electrolyte molecules are ionized. In this chapter, substances classified as weak acids, weak bases, and salts are extensively treated from a chemical equilibrium point of view. Involved in the discussion are these concepts: buffers, hydrolysis, pH, pOH, and reactions involving weak acids and weak bases.

The problems emphasize equilibria of weak acids, weak bases, and salts in aqueous solutions. Calculations involving pH and pOH require logarithms. Part II of this book, Exponential Notation and Logarithms, is helpful for reviewing logarithm concepts.

FORMULAS AND DEFINITIONS

Buffer solution Solution containing the salt of a weak acid or weak base along with the weak acid or weak base, respectively.

Ionization constant (K_i) Equilibrium constant describing the ionization of a weak acid or weak base in water solution.

Ion-product constant for water (K_w) Equilibrium constant describing the relation between the hydrogen ion and the hydroxyl ion concentrations in aqueous solution:

$$K_w = [H^+] [OH^-] = 10^{-14} \text{ at } 25°C$$

Hydrolysis Reaction of a salt of a weak acid or weak base with water.

Hydrolysis constant (K_h) Equilibrium constant describing the hydrolysis of a salt in aqueous solution.

pH Measure of hydrogen ion concentration in a solution; defined as $pH = -\log [H^+]$.

PROBLEMS

1. Calculate the concentration of each of the ions in the following solutions:
   (a) 0.03 M HCl; (b) 0.002 M Ca(OH)$_2$.

   Soln: HCl and Ca(OH)$_2$ are strong electrolytes and, it can be assumed, completely ionized in solution.

   (a) 0.03 M HCl produces 0.03 M H$^+$ and 0.03 M Cl$^-$. The concentration of OH$^-$ is governed by the equilibrium, $H_2O \rightleftharpoons H^+ + OH^-$, where $K_w = [H^+] [OH^-] = 1 \times 10^{-14}$. Substitution gives

   $$[0.03] [OH^-] = 1 \times 10^{-14}$$

   $$[OH^-] = \frac{1 \times 10^{-14}}{3 \times 10^{-2}} = 3 \times 10^{-13}$$

   (b) 0.002 M Ca(OH)$_2$ produces 0.002 M Ca^{2+} and 0.004 M OH$^-$. The concentration of H$^+$ is governed by the equilibrium $H_2O \rightleftharpoons H^+ + OH^-$, where

   $$K_w = [H^+] [OH^-] = 1 \times 10^{-14}$$

   Substitution gives

   $$[H^+] [0.004] = 1 \times 10^{-14}$$

   $$[H^+] = \frac{1 \times 10^{-14}}{4 \times 10^{-3}} = 2.5 \times 10^{-12}$$

2. The ionization constant of HClO is 3.5×10^{-8}. Calculate [H$^+$], [ClO$^-$], and the percent ionization of a 0.050 M solution of the acid.

   Soln: The ionization of HClO is HClO $\rightleftharpoons$ H$^+$ + ClO$^-$. $K_i = 3.5 \times 10^{-8}$. Let x equal the number of moles of HClO ionizing. Then $x = [H^+] = [ClO^-]$ and $0.05 - x = [HClO]$ at equilibrium:

   $$\frac{(x)(x)}{(0.050 - x)} = 3.5 \times 10^{-8}$$

   For the factor $0.050 - x$, x is very small relative to 0.050 and can be neglected, which greatly simplifies the calculation. In general, if the initial concentration of the ionizing substance and K_i for the substances differ by an exponential factor of 3 or more, x can be neglected. In this case, the concentration of HClO is 0.050, or 5.0×10^{-2} M, and K_i is 3.5×10^{-8}. The exponents differ by 6; hence, x can be neglected in the denominator. Likewise, if the exponential difference is less that 3, x cannot be ignored and the value of x must be determined by using the quadratic formula. When x is ignored, the equation becomes

$$\frac{x^2}{0.050} = 3.5 \times 10^{-8}$$

or

$$x^2 = 1.75 \times 10^{-9}$$

$$x = 4.2 \times 10^{-5} = [H^+] = [ClO^-]$$

$$\% \text{ ionization} = \frac{4.2 \times 10^{-5}}{0.050} \times 100 = 8.4 \times 10^{-2}\%$$

3. A 0.010 M solution of HNO_2 is 19 percent ionized. Calculate $[H^+]$, $[NO_2^-]$, and $[HNO_2]$, and the ionization constant for HNO_2.

Soln: 19% of the HNO_2 is ionized in solution: $HNO_2 \rightleftharpoons H^+ + NO_2^-$.

$$[H^+] = [NO_2^-] = (0.19)(0.010\ M)$$

$$[HNO_2] = (0.010 - 0.19(0.010))M$$

$$\frac{[H^+][NO_2^-]}{[HNO_2]} = K_i = \frac{(1.9 \times 10^{-3})^2}{(8.1 \times 10^{-3})}$$

$$= \frac{3.61 \times 10^{-6}}{8.1 \times 10^{-3}} = 4.5 \times 10^{-4}$$

4. Calculate the percent ionization of each of the following solutions. Calculate (1) using simplifying assumption and (2) without using simplifying assumption and with use of the quadratic formula. Compare the values.
(a) 0.020 M HCOOH. $K_i = 1.8 \times 10^{-4}$.

Soln: (a) $HCOOH \rightleftharpoons H^+ + HCOO^-$; $K_i = \dfrac{[H^+][HCOO^-]}{[HCOOH]} = 1.8 \times 10^{-4}$

Let x equal the number of moles of HCOOH ionizing.

$$x = [H^+] = [HCOO^-] \quad \text{and} \quad 0.020 - x = [HCOOH]$$

$$\frac{(x)(x)}{(0.020 - x)} = 1.8 \times 10^{-4}$$

With simplification,

$$x^2 = (1.8 \times 10^{-4})(0.020) = 3.6 \times 10^{-6}$$

$$x = 1.9 \times 10^{-3}$$

$$\% \text{ ionization} = \frac{1.9 \times 10^{-3}}{0.020} \times 100 = 9.5\%$$

Without simplification,

$$\frac{x^2}{0.020 - x} = 1.8 \times 10^{-4}$$

$$x^2 = 3.6 \times 10^{-6} - (1.8 \times 10^{-4})x$$

Transposing terms gives

$$x^2 + (1.8 \times 10^{-4})x - 3.6 \times 10^{-6} = 0$$

Solve for x via the quadratic formula:

$$x = \frac{-1.8 \times 10^{-4} \pm \sqrt{(1.8 \times 10^{-4})^2 - 4(1)(-3.6 \times 10^{-6})}}{2}$$

$$= \frac{-1.8 \times 10^{-4} \pm \sqrt{3.24 \times 10^{-8} + 1.44 \times 10^{-5}}}{2}$$

In this case, 3.24×10^{-8} does not add significantly to 1.44×10^{-5} and is neglected:

$$x = \frac{-1.8 \times 10^{-4} \pm \sqrt{3.80 \times 10^{-3}}}{2}$$

The roots are -1.99×10^{-5} and 1.8×10^{-3}.

$$\% \text{ ionization} = \frac{1.8 \times 10^{-3}}{0.020} \times 100 = 9.0\%$$

For this case 9.0% is significantly different from 9.5%. Moreover, the simplification rule would not allow neglecting x!

5. The ionization constant of triethylamine, $(C_2H_5)_3N$, is 7.40×10^{-5}.

$$(C_2H_5)_3N + H_2O \rightleftharpoons (C_2H_5)_3NH^+ + OH^-$$

Calculate the hydroxyl ion concentration in a 1.50 M triethylamine solution.

Soln: Triethylamine ionizes as $(C_2H_5)_3N + H_2O \rightleftharpoons (C_2H_5)_3NH^+ + OH^-$ and

$$K_i = \frac{[(C_2H_5)_3NH^+][OH^-]}{[(C_2H_5)_3N]} = 7.40 \times 10^{-5}$$

Let x equal the number of moles of the amine ionizing. Then $x = [(C_2H_5)_3NH^+] = [OH^-]$ and $1.50 - x = [(C_2H_5)_3N]$ at equilibrium.

$$\frac{(x)\,(x)}{(1.50 - x)} = 7.40 \times 10^{-5}$$

Is x small relative to $1.50\,M$? Yes.

$$\frac{x^2}{1.50} = 7.40 \times 10^{-5} \quad \text{and} \quad x^2 = 11.1 \times 10^{-5}$$

$$x = 1.05 \times 10^{-2} = [OH^-]$$

6. Calculate the molarity of a solution of acetic acid which is 2.0% ionized. ($K_i = 1.8 \times 10^{-5}$.)

Soln: The problem states that 2 percent of the number of moles of molecular acetic acid (abbreviated HOAc for this and the remainder of the problems) are dissociated into ions. HOAc $\rightleftharpoons$ H$^+$ + OAc$^-$; $K_i = 1.8 \times 10^{-5}$.

$$K_i = \frac{[H^+]\,[OAc^-]}{[HOAc]} = 1.8 \times 10^{-5}$$

Let x equal the initial molarity of HOAc. Then $0.02x = [H^+] = [OAc^-]$ and $x - 0.02x = [HOAc]$ at equilibrium.

$$\frac{(0.02x)(0.02x)}{x - 0.02x} = 1.8 \times 10^{-5}$$

$$(4 \times 10^{-4})x^2 = (1.8 \times 10^{-5})\,(0.98x) = (1.76 \times 10^{-5})x$$

Dividing both sides by $(4 \times 10^{-4})x$ yields

$$x = \frac{1.76 \times 10^{-5}}{4 \times 10^{-4}} = 0.044\,M = [HOAc]$$

12. Calculate the pH and $[OH^-]$ for a $0.23\,M$ cyclohexylamine, $C_6H_{11}NH_2$, solution. (K_i for cyclohexylamine is 4.6×10^{-4}.)

Soln: The compound is ionized in water as

$$C_6H_{11}NH_2 + H_2O \rightleftharpoons C_6H_{11}NH_3^+ + OH^-$$

$$K_i = \frac{[C_6H_{11}NH_3^+]\,[OH^-]}{[C_6H_{11}NH_2]} = 4.6 \times 10^{-4}$$

Let x equal the number of moles of $C_6H_{11}NH_2$ that ionize. Then $x = [C_6H_{11}NH_3^+] = [OH^-]$ and $0.23 - x = [C_6H_{11}NH_2]$.

$$\frac{(x)\,(x)}{(0.23 - x)} = 4.6 \times 10^{-4}$$

Is x small relative to 0.23? Barely! Multiply through and solve for x:

$$x^2 = 1.06 \times 10^{-4}$$

$$x = 0.010\,M = [OH^-]$$

The $[H^+]$ is calculated from $[\hat{H}^+][OH^-] = 10^{-14}$:

$$[H^+] = \frac{10^{-14}}{[OH^-]} = \frac{10^{-14}}{10^{-2}} = 10^{-12}$$

$$pH = -\log[H^+] = -\log 10^{-12} = -(-12) = 12$$

16. The ionization constant of acetic acid is 1.8×10^{-5}. What will be the pH of a 0.10 M HOAc solution which is 0.10 M in NaOAc?

Soln: This solution is a buffer solution containing a weak acid, HOAc, and its salt, NaOAc. The acetic acid equilibrium describes the solution, since the added NaOAc merely produces a shift in the equilibrium. HOAc $\rightleftharpoons H^+ + OAc^-$;

$$K_i = \frac{[H^+][OAc^-]}{[HOAc]} = 1.8 \times 10^{-5}$$

Let x equal the number of moles of HOAc ionizing. Then $x = [H^+]$; $[OAc^-] = (0.10 + x)\, M$.

$$[HOAc] = (0.10 - x)\, M$$

$$\frac{(x)(0.10 + x)}{(0.10 - x)} = 1.8 \times 10^{-5}$$

Test the simplification procedure: x is small relative to 0.10 and can be ignored in both cases.

$$\frac{(x)(0.10)}{(0.10)} = 1.8 \times 10^{-5}$$

$$x = 1.8 \times 10^{-5} = [H^+]$$

The pH is calculated as

$$pH = -\log[H^+]$$
$$= -\log 1.8 \times 10^{-5}$$
$$= -[\log 1.8 + \log 10^{-5}]$$
$$= -[0.2553 + (-5)]$$
$$= -(-4.74)$$
$$= 4.74$$

17. The ionization constant for acetic acid is 1.8×10^{-5}. How many grams of $NaC_2H_3O_2$ must be added to one liter of a 0.20 M solution of $HC_2H_3O_2$ to maintain a hydrogen ion concentration of 6.5×10^{-5} M?

Soln: As in Problem 16, this solution is a buffer solution and the acetic acid equilibrium applies. Since the equilibrium is defined by molarity, the concentration of the salt must be calculated and then converted to weight.

$$1.8 \times 10^{-5} = \frac{[H^+] [OAc^-]}{[HOAc]}$$

Let x equal the concentration of OAc^- required:

$$1.8 \times 10^{-5} = \frac{(6.5 \times 10^{-5}) (x)}{(0.20 - 6.5 \times 10^{-5})}$$

Simplifying gives

$$x = \frac{(0.20) (1.8 \times 10^{-5})}{(6.5 \times 10^{-5})} = 5.54 \times 10^{-2} \; M$$

Therefore, we can obtain the weight of $NaC_2 H_3 O_2$:

$$Wt = (5.54 \times 10^{-2} \; \text{mole}) (82.0 \; \text{g/mole}) = 4.5 \; \text{g}$$

20. Calculate the hydrogen ion concentration for a solution having a pOH of 6.34.

Soln: The pOH of the solution is 6.34. Therefore,

$$pH = 14.00 - pOH = 14.00 - 6.34 = 7.66$$

Since $pH = -\log[H^+]$,

$$[H^+] = 10^{-pH} = 10^{-7.66} = 10^{0.34} \times 10^{-8}$$

The antilog of 0.34 is 2.2, and $[H^+]$ is 2.2×10^{-8}.

22. What is the pH of a solution containing aqueous ammonia and ammonium chloride having (a) a molar ratio of 5.0 to 3.0?

Soln: This is a buffer solution of a weak base, NH_3, and its salt, $NH_4 Cl$. The NH_3 equilibrium describes the system.
(a) $NH_3 + H_2 O \rightleftharpoons NH_4^+ + OH^-$

$$K_i = 1.8 \times 10^{-5} = \frac{[NH_4^+] [OH^-]}{[NH_3]}$$

The ratio of $[NH_3]$ to $[NH_4^+]$ is 5.0 to 3.0.

$$1.8 \times 10^{-5} = \frac{(3.0) [OH^-]}{(5.0)}$$

$$[OH^-] = 3 \times 10^{-5}$$

$$pOH = -\log 3 \times 10^{-5}$$

$$= -[\log 3 + \log 10^{-5}]$$

$$= -[0.4771 + (-5)]$$

$$= 4.52$$

$$pH = 9.48$$

25. A 25-ml volume of 0.10 M lactic acid is added to 5.0 ml of 0.69 M sodium lactate solution. Calculate the pH of this solution ($K_i = 1.37 \times 10^{-4}$).

Soln: Buffer solution of lactic acid–sodium lactate.

$$K_i = 1.37 \times 10^{-4} = \frac{[H^+] \ [lactate^-]}{[lactic \ acid]}$$

Initial concentrations are

$$L. \ acid = (25 \ ml)(0.10 \ M) = \frac{2.5 \ m\text{-}moles}{30 \ ml}$$

$$Na \ lactate = (5.0 \ ml) \ (0.69 \ M) = \frac{3.45 \ m\text{-}moles}{30 \ ml}$$

Let x equal the number of m-moles of lactic acid that ionize:

$$1.37 \times 10^{-4} = \frac{(x) \ [(3.45 \ m\text{-}moles + x)/30 \ ml]}{(2.5 \ m\text{-}moles - x)/30 \ ml}$$

Simplifying and transposing terms gives:

$$\frac{(2.5) \ (1.37 \times 10^{-4})}{3.45} = x = 9.9 \times 10^{-5} = [H^+]$$

$$pH = -\log 9.9 \times 10^{-5}$$
$$= -(\log 9.9 + \log 10^{-5})$$
$$= -(\sim 1 + (-5))$$
$$= 4.0$$

26. The pH of a given HF solution is 1.57. What is the percent ionization of the acid? ($K_i = 7.2 \times 10^{-4}$.)

Soln: The pH of the solution is 1.57. What must be the initial [HF]?
$$HF \rightleftharpoons H^+ + F^-. \ K_i = 7.2 \times 10^{-4}.$$

$$[H^+] = [F^-] = 10^{-pH} = 10^{-1.57} = 10^{0.43} \times 10^{-2} = 2.69 \times 10^{-2}$$

Let x equal the initial [HF]:

$$\frac{(2.69 \times 10^{-2})^2}{x - 2.69 \times 10^{-2}} = 7.2 \times 10^{-4}$$

$$7.24 \times 10^{-4} = (7.2 \times 10^{-4}) \ x - 1.94 \times 10^{-5}$$

Transposing terms and dividing by 7.2×10^{-4} gives $x = 1.03 \ M =$ [HF].

$$\% \ ionization = \frac{2.69 \times 10^{-2}}{1.03} \times 100 = 2.6\%$$

27. Forty ml of 0.20 M HCl is mixed with 50 ml of 0.30 M aqueous ammonia. Calculate the concentration of hydroxyl ions in the solution.

Soln: The reaction of HCl with NH_3 is $NH_3 + HCl \rightarrow NH_4Cl$. Consider the amount of each reagent used and decide which reagent is in excess. Initial quantities are

HCl   No. m-moles = (40 ml) (0.20 M) = 8.0

NH_3 No. m-moles = (50 ml) (0.30 M) = 15.0

According to the reaction, 8.0 m-moles of HCl will consume 8.0 m-moles of NH_3 and leave 7.0 m-moles. The resulting solution is a buffer of NH_3 and NH_4^+:

$$NH_3 + H_2O \rightleftharpoons NH_4^+ + OH^-$$

$$K_i = 1.8 \times 10^{-5} = \frac{[OH^-](8.0 \text{ m-moles}/90 \text{ ml})}{(7.0 \text{ m-moles}/90 \text{ ml})}$$

$$[OH^-] = 1.6 \times 10^{-5}$$

30. (a) In what mole ratio would you combine sodium acetate and acetic acid in water solution to prepare a buffer solution of pH 7.00?

Soln: (a) pH = 7.00. The buffer solution is described by the HOAc equilibrium, $HOAc \rightleftharpoons H^+ + OAc^-$. For a pH of 7.00, $[H^+] = 10^{-7}$. The ratio of OAc^- to HOAc is calculated from the equilibrium expression:

$$1.8 \times 10^{-5} = \frac{[H^+] [OAc^-]}{[HOAc]} = (10^{-7}) \frac{[OAc^-]}{[HOAc]}$$

$$\frac{[OAc^-]}{[HOAc]} = 1.8 \times 10^2$$

$$[OAc^-] = 180[HOAc]$$

32. Calculate the molarity of a solution of H_2SO_4 which has a pH of 2.00. (Be careful on this one. It is more difficult than it looks. Remember that H_2SO_4 is a diprotic acid.)

Soln: The ionization of H_2SO_4 occurs in two steps:

$$H_2SO_4 \rightarrow H^+ + HSO_4^-$$
$$HSO_4^- \rightleftharpoons H^+ + SO_4^{2-}$$

$$K_i = 1.2 \times 10^{-2}$$

Total $[H^+]$ is the sum of $[H^+]$ arising from the two steps. Since the pH of the solution is to be 2.0, total $[H^+]$ equals 10^{-2}, or 0.01 M. Actually, two quantities are unknown, $[H^+]$ from H_2SO_4, Step 1, and $[H^+]$ from HSO_4^-, Step 2.

Let x = $[H^+]$ from HSO_4^- = $[SO_4^{2-}]$, and y = $[H^+]$ from H_2SO_4 and the initial $[HSO_4^-]$. $x + y = 0.0100$ or $y = 0.0100 - x$.

For the second ionization,

$$\frac{[H^+]\ [SO_4{}^{2-}]}{[HSO_4{}^-]} = 1.2 \times 10^{-2}$$

Substituting x and y into the equation gives

$$\frac{(x + y)\ (x)}{(y - x)} = 1.2 \times 10^{-2}$$

But $y = 0.0100 - x$

$$\frac{(x + (0.0100 - x))(x)}{(0.0100 - x) - x} = 1.2 \times 10^{-2}$$

Simplifying gives

$$\frac{0.0100x}{0.0100 - 2x} = 1.2 \times 10^{-2}$$

$$x = 0.0035$$

Initial $[H_2 SO_4] = 0.0100 - 0.0035 = 0.0065\ M.$

38. A 50.0-ml solution of a 0.10 M monoprotic acid was mixed with 20 ml of 0.10 M potassium hydroxide, and the resulting mixture was diluted to 100 ml. The pH of the solution was found to be 5.25. What is the ionization constant of the acid?

Soln: The reaction between KOH and HA is $KOH + HA \rightarrow KA + H_2 O$. Initial amounts of reagent are

HA   No. m-moles = (50 ml) (0.10 M) = 5.00

KOH   No. m-moles = (20.0 ml) (0.10 M) = 2.00

The reaction consumes all the KOH, leaves 3.00 m-moles of HA unreacted, and produces 2.00 m-moles of KA. For a pH of 5.25, $[H^+]$ is 5.62×10^{-6}. Substituting the values into the expression for K_i gives

$$K_i = \frac{[H^+]\ [A^-]}{[HA]}$$

$$= \frac{(5.62 \times 10^{-6})(2.00\ \text{m-moles}/100\ \text{ml})}{3.00\ \text{m-moles}/100\ \text{ml}} = 3.7 \times 10^{-6}$$

43. Calculate the concentration of sulfide ion in a saturated solution of hydrogen sulfide which is 0.3 M with respect to hydrochloric acid. ($K_{H_2S} = 1 \times 10^{-7}$, and $K_{HS^-} = 1 \times 10^{-13}$.)

Soln: The ionization of $H_2 S$ occurs in two steps:

$$H_2 S \rightleftharpoons H^+ + HS^-; \qquad K_1 = 1 \times 10^{-7}$$

$$HS^- \rightleftharpoons H^+ + S^{2-}; \qquad K_2 = 1 \times 10^{-13}$$

An equilibrium expression containing S^{2-} is derived by multiplying the expression for K_1 and the expression for K_2:

$$\frac{[H^+]\ [HS^-]}{[H_2 S]} \times \frac{[H^+]\ [S^{2-}]}{[HS^-]} = 1 \times 10^{-20}$$

Dividing out terms gives

$$\frac{[H^+]^2\ [S^{2-}]}{[H_2 S]} = 1 \times 10^{-20}$$

A saturated $H_2 S$ solution is approximately $0.10\ M$. The $[H^+]$ in an $H_2 S$ solution that is $0.30\ M$ with respect to HCl, will be essentially equal to the concentration of the HCl. Therefore, $[S^{2-}]$ is

$$\frac{(0.30)^2\ [S^{2-}]}{(0.10)} = 1 \times 10^{-20}$$

$$[S^{2-}] = 1 \times 10^{-20} \qquad \text{(one significant figure)}$$

47. Calculate the pH of a $0.0050\ M$ solution of hydrogen sulfide. (Neglect the second stage in the ionization of $H_2 S$.) $K_{H_2S} = 1 \times 10^{-7}$.

Soln: The first ionization of $H_2 S$ is $H_2 S \rightleftharpoons H^+ + HS^-$; $K_1 = 1 \times 10^{-7}$. Let x equal the number of moles of $H_2 S$ ionizing. $x = [H^+] = [HS^-]$ and $[H_2 S] = 0.0050 - x$.

$$\frac{(x)\ (x)}{(0.0050 - x)} = 1 \times 10^{-7}$$

Simplify and solve for x, obtaining $x = 2.24 \times 10^{-5}$. The $pH = 4.65$.

49. Calculate (a) the carbonate ion concentration and (b) the bicarbonate ion concentration in a $0.050\ M$ solution of carbonic acid which is also $0.10\ M$ in hydrochloric acid. ($K_1 = 4.3 \times 10^{-7}$, and $K_2 = 7 \times 10^{-11}$ for the two steps in the dissociation of $H_2 CO_3$.)

Soln: The $[H^+]$ effectively equals [HCl] since the first ionization of $H_2 CO_3$ is very slight and the second is negligible.

$$H_2 CO_3 \rightleftharpoons H^+ + HCO_3^-; \qquad K_1 = 4.3 \times 10^{-7}$$
$$HCO_3^- \rightleftharpoons H^+ + CO_3^{2-}; \qquad K_2 = 7 \times 10^{-11}$$

(a) Multiply both equilibrium expressions together to get an expression containing CO_3^{2-}.

$$\frac{[H^+]^2\ [CO_3^{2-}]}{[H_2 CO_3]} = 3.01 \times 10^{-17}$$

Since $[H^+] = [HCl] = 0.10$,

$$[CO_3{}^{2-}] = \frac{(0.10)^2 [CO_3{}^{2-}]}{(0.050)} = 3.01 \times 10^{-17}$$

$$[CO_3{}^{2-}] = 1.5 \times 10^{-16} \text{ or } 2 \times 10^{-16}$$

(b) Similarly for $[HCO_3{}^-]$:

$$\frac{[H^+][HCO_3{}^-]}{[H_2 CO_3]} = 4.3 \times 10^{-7}$$

$$[HCO_3{}^-] = 2.2 \times 10^{-7}$$

51. A 0.100-liter volume of 0.010 M aqueous ammonia solution is placed in a chamber exposed to HCl gas. Calculate the pH of the solution for each of the following situations (you may assume that dissolved HCl does not increase the volume of the solution significantly):
(a) Before any HCl dissolves in the ammonia solution.
(b) After 1.0×10^{-4} mole of HCl has dissolved.
(c) After 5.0×10^{-4} mole of HCl has dissolved.
(d) After 9.0×10^{-4} mole HCl has dissolved.
(e) After 1.0×10^{-3} mole HCl has dissolved.

Soln: This example is essentially the titration of ammonia against hydrochloric acid. In part (a) the solution contains only ammonia. Solve the ammonia equilibrium expression for $[OH^-]$, then calculate pH. In parts b, c, and d, the solution contains $NH_4{}^+$ along with NH_3, a buffer solution. However, in part (e) all the NH_3 has been allowed to react and the solution contains only $NH_4 Cl$, which hydrolyzes to $NH_4{}^+ + H_2 O \rightleftharpoons NH_3 + H_3 O^+$.

(b) Initial amounts are: number of moles of NH_3 = (0.100 liter) (0.010 M) = 0.0010 mole. Adding 0.0001 mole of HCl consumes 0.0001 mole of NH_3 and produces 0.0001 mole of $NH_4{}^+$. Therefore, the NH_3 equilibrium describes the solution. Let x equal the number of moles of NH_3 ionizing.

$$x = [OH^-]; \quad [NH_4{}^+] = 0.001 + x; \quad [NH_3] = 0.009 - x$$

$$K_i = \frac{[NH_4{}^+][OH^-]}{[NH_3]} = 1.8 \times 10^{-5}$$

$$\frac{(0.0010 + x)(x)}{(0.009 - x)} = 1.8 \times 10^{-5}$$

The value of x can be considered small relative to 0.0010 M and 0.009 M. The expression reduces to

$$\frac{0.0010x}{0.009} = 1.8 \times 10^{-5}$$

$$x = 1.62 \times 10^{-4}$$

$p\text{OH} = 3.79, p\text{H} = 10.21.$

(c) The ratio of NH_4^+ to NH_3 is 1 to 1 and $[OH^-] = K_i = 1.8 \times 10^{-5}$, $p\text{H} = 9.26$.

(d) Same as part (b), except that $[NH_4^+]$ and $[NH_3]$ are reversed. The expression must be solved via the quadratic formula.

(e) Equal molar quantities of HCl and NH_3 have been allowed to react and the solution now is a 0.010 M NH_4Cl solution. The NH_4^+ hydrolyzes to $NH_4^+ + H_2O \rightleftharpoons NH_3 + H_3O$; $K_h = 5.5 \times 10^{-10}$. Let x equal the number of moles of NH_4^+ hydrolyzing. $x = [NH_3] = [H_3O^+]$; $[NH_4^+] = 0.001 - x$.

$$\frac{(x)(x)}{(0.01 - x)} = 5.5 \times 10^{-10}$$

Simplify and solve for x:

$$x = [H_3O^+] = 2.35 \times 10^{-6}$$

$$p\text{H} = 5.63$$

57. Assuming that the hydrogen ion concentration in the solution is controlled by the reaction $H_2AsO_4^- \rightleftharpoons H^+ + HAsO_4^{2-}$, calculate the ratio of concentrations of $HAsO_4^{2-}$ necessary to make up a buffer solution of $p\text{H}$ equal to (a) 6.00, (b) 7.20, (c) 8.00. (K_i for the reaction $H_2AsO_4^- \rightleftharpoons H^+ + HAsO_4^{2-}$ is 1.0×10^{-7}.)

Soln: (a) The ratio of concentrations is calculated from the equilibrium expression

$$K_i = 1.0 \times 10^{-7} = \frac{[H^+][HAsO_4^{2-}]}{[H_2AsO_4^-]}$$

The $p\text{H}$ of the buffer is to be 6.00. Therefore,

$$\frac{[HAsO_4^{2-}]}{[H_2AsO_4^-]} = \frac{1.0 \times 10^{-7}}{[H^+]} = \frac{1.0 \times 10^{-7}}{10^{-6}} = 0.1$$

58. Calculate the $p\text{H}$ of a 0.10 M sodium acetate solution.

Soln: Salts, such as sodium acetate, are considered to be completely dissociated in solution. The hydrolysis of NaOAc is given by $OAc^- + H_2O \rightleftharpoons HOAc + OH^-$.

The hydrolysis constant concept is treated in Sections 18.17 and 18.18 of the textbook. For this example,

$$K_h = \frac{[HOAc][OH^-]}{[OAc^-]} \equiv \frac{[HOAc] K_w}{[OAc^-][H^+]} = \frac{K_w}{K_{i(HOAC)}}$$
$$= \frac{1 \times 10^{-14}}{1.8 \times 10^{-5}} = 5.5 \times 10^{-10}$$

Let x equal the number of moles of OAc^- that are hydrolyzed. $x = [HOAc] = [OH^-]$ and $0.10 - x = [OAc^-]$.

$$\frac{(x)\,(x)}{0.10 - x} = 5.5 \times 10^{-10}$$

Simplifying and solving for x yields

$$x = 7.42 \times 10^{-6} = [OH^-]$$

$$pH = 8.87$$

59. Calculate the hydrolysis constant for each of the following ions:
 (a) $CH_3CH_2COO^-$ (K_i for $CH_3CH_2COOH = 1.3 \times 10^{-5}$).
 (b) ClO^- (K_i for $HClO = 3.5 \times 10^{-8}$).
 (c) N_3^- (K_i for $HN_3 = 1 \times 10^{-4}$).
 (d) $C_2O_4^{2-}$ (K_i for $HC_2O_4^- = 6.4 \times 10^{-5}$).

Soln: The hydrolysis constant for the ion of a weak acid or a weak base is K_w/K_i for the acid or base.

 (a) $CH_3CH_2COO^- + H_2O \rightleftharpoons CH_3CH_2COOH + OH^-$

$$K_i = \frac{K_w}{K_h} = \frac{1 \times 10^{-14}}{1.3 \times 10^{-5}} = 7.7 \times 10^{-10}$$

60. Calculate the pH of (a) 0.500 M solution of ammonium chloride; (b) a 0.040 M solution of NaF. (K_i for NH_3 is 1.8×10^{-5}, and K_i for HF is 7.2×10^{-4}.)

Soln: (a) NH_4Cl hydrolyzes to $NH_4^+ + H_2O \rightleftharpoons NH_3 + H_3O^+$.

$$K_h = \frac{K_w}{K_{i(NH_3)}} = \frac{1 \times 10^{-14}}{1.8 \times 10^{-5}} = 5.5 \times 10^{-10}$$

Let x equal the number of moles of NH_4^+ that hydrolyze. $x = [NH_3] = [H_3O^+]$ and $0.500 - x = [NH_4^+]$.

$$\frac{(x)\,(x)}{0.500 - x} = 5.5 \times 10^{-10}$$

Simplifying and solving for x yields

$$x = 1.66 \times 10^{-5} = [H_3O^+]$$

$$pH = 4.78$$

63. What will be the final pH of a solution made by mixing 0.20 mole of NaOH and 0.20 mole of NH_4NO_3 in enough water to make 1.0 liter of solution?

Soln: The reaction between NH_4NO_3 and NaOH is an acid-base reaction as shown below:

$$NH_4NO_3 + H_2O \rightarrow NH_4^+ + NO_3^-$$

$$NH_4^+ + H_2O \rightleftharpoons NH_3 + H_3O^+$$

$$NaOH + H_3O^+ \rightarrow Na^+ + 2H_2O$$

The reaction between NH_4^+ and NaOH occurs on a one mole–one mole basis. Since equimolar amounts of NH_4NO_3 and NaOH are used, the resulting solution is essentially a 0.20 M NH_3 solution described by the ammonia equilibrium $NH_3 + H_2O \rightleftharpoons NH_4^+ + OH^-$.

$$K_i = 1.8 \times 10^{-5} = \frac{[NH_4^+] \, [OH^-]}{[NH_3]}$$

Let x equal the number of moles of ammonia that are ionized. $x = [OH^-] = [NH_4^+]$ and $0.20 - x = [NH_3]$.

$$\frac{(x) \, (x)}{(0.20 - x)} = 1.8 \times 10^{-5}$$

Solving for x gives

$$x = 1.9 \times 10^{-3} = [OH^-]$$

$$pOH = 2.72 \qquad pH = 11.28$$

65. Calculate the pH of a 0.350 M Na_2HAsO_4 solution. (Neglect the dissociation of the $HAsO_4^{2-}$ ion.) (K_i for $H_2AsO_4^-$ is 1.0×10^{-7}.)

Soln: The hydrolysis of the $HAsO_4^{2-}$ ion occurs as $HAsO_4^{2-} + H_2O \rightleftharpoons H_2AsO_4^- + OH^-$.

$$K_h = \frac{K_w}{K_i} = \frac{1.0 \times 10^{-14}}{1.0 \times 10^{-7}} = \frac{[H_2AsO_4^-] \, [OH^-]}{[HAsO_4^{2-}]}$$

Let x equal the number of moles of $HAsO_4^{2-}$ hydrolyzing. $x = [H_2AsO_4^-] = [OH^-]$ and $[H_2AsO_4^{2-}] = 0.350 - x$.

$$\frac{(x) \, (x)}{0.350 - x} = 1.0 \times 10^{-7}$$

Simplifying and solving for x gives

$$x = 1.87 \times 10^{-4} = [OH^-]$$

$$pH = 10.27$$

70. Calculate the pH of a solution made by mixing equal volumes of 0.040 M aqueous aniline, $C_6H_5NH_2$, and 0.040 M nitric acid.

Soln: Aniline is a weak base and ionizes as $C_6H_5NH_2 + H_2O \rightleftharpoons C_6H_5NH_3^+ + OH^-$. After reaction, the resulting solution contains $C_6H_5NH_3^+$ in a volume equal to the sum of the solution volumes used. To simplify calculations, assume that 1.0 liter of each solution is used.

Initial $[C_6H_5NH_3^+] = \dfrac{(0.040\ M)\ (1.0\ \text{liter})}{2.0\ \text{liters}} = 0.020\ M$

$$C_6H_5NH_3^+ + H_2O \rightleftharpoons C_2H_5NH_2 + H_3O^+$$

Let x equal the number of moles of $C_6H_5NH_3^+$ hydrolyzing: $x = [C_6H_5NH_2] = [H_3O^+]$ and $0.020 - x = [C_6H_5NH_3^+]$.

$$\frac{(x)\ (x)}{0.02 - x} = K_h = \frac{K_w}{K_i} = 2.17 \times 10^{-5}$$

Simplifying and solving for x yields

$$x = 6.59 \times 10^{-4} = [H_3O^+]$$

$$pH = 3.18$$

72. How many moles of $NaC_2H_3O_2$ must be used to prepare 2.0 liters of aqueous solution with $pH = 9.40$? (K_i for $HC_2H_3O_2 = 1.8 \times 10^{-5}$.)

Soln: The hydrolysis of $NaC_2H_3O_2$ (NaOAc) gives $C_2H_3O_2^- + H_2O \rightleftharpoons HC_2H_3O_2 + OH^-$.

$$K_h = 5.5 \times 10^{-10} = \frac{[HOAc]\ [OH^-]}{[OAc^-]}$$

The pH of the solution is 9.40. Therefore, the pOH is 4.6 and the $[OH^-]$ is 2.51×10^{-5}. Let x equal the initial concentration of OAc^-. $[HOAc] = [OH^-] = 2.51 \times 10^{-5}$ and $[OAc^-] = x - 2.51 \times 10^{-5}$

$$\frac{(2.51 \times 10^{-5})^2}{x - 2.51 \times 10^{-5}} = 5.5 \times 10^{-10}$$

The quantity x is large compared with 2.51×10^{-5}, and the equation can be simplified to

$$\frac{(2.51 \times 10^{-5})^2}{x} = 5.5 \times 10^{-10}$$

Solving for x gives $x = 1.15$ moles/liter $= [OAc^-]$. Since 2.0 liters are to be prepared, 2(1.15 moles), or 2.3 moles, are required.

74. Calculate the pH of each of the solutions described:
    (a) 40 ml of 0.10 M barbituric acid ($K_i = 9.8 \times 10^{-5}$).
    (b) 40 ml of 0.10 M barbituric acid and 20 ml of 0.10 M KOH.
    (c) 40 ml of 0.10 M barbituric acid and 39 ml of 0.10 M KOH.
    (d) 40 ml of 0.10 M barbituric acid and 40 ml of 0.10 M KOH.
    (e) 40 ml of 0.10 M barbituric acid and 41 ml of 0.10 M KOH.

Soln: For brevity, let HB represent barbituric acid and B^- respresent the barbiturate ion. This example involves titrating HB against KOH, beginning with the acid solution and titrating through the end point to a hydrolytic solution containing B^- and finally to a solution containing excess OH^-. The acid ionizes as $HB + H_2O \rightleftharpoons H_3O^+ + B^-$; $K_i = 9.8 \times 10^{-5}$.

(a) Let x equal the number of moles of HB ionizing. $x = [H_3O^+] = [B^-]$ and $0.10 - x = [HB]$.

$$\frac{(x)\,(x)}{0.10 - x} = 9.8 \times 10^{-5}$$

$$x = 3.13 \times 10^{-3} = [H_3O^+]$$

$$pH = 2.51$$

(b) One-half of the HB is consumed. Therefore, $[B^-] = [HB]$ remaining.

$$K_i = \frac{[H_3O^+]\,[B^-]}{[HB]}$$

$$= [H_3O^+] = 9.8 \times 10^{-5}$$

$$pH = 4.01$$

(c) Only 1.0 ml of HB remains. The ratio of $[B^-]$ to $[HB]$ is:

No. m-moles of HB $= (1.0 \text{ ml}) (0.10\, M) = 0.10$

No. m-moles of $B^- = (39 \text{ ml}) (0.10\, M) = 3.9$

$$K_i = 9.8 \times 10^{-5} = \frac{[H_3O^+]\,[B^-]}{[HB]} = \frac{[H_3O^+]\,(3.9 \text{ m-moles})}{(0.10 \text{ m-mole})}$$

$$[H_3O^+] = 2.51 \times 10^{-6}$$

$$pH = 5.60$$

(d) All the HB is consumed. Therefore, a hydrolytic solution of B^- exists. The B^- produced is contained in 80 ml of solution.

No. m-moles B^- produced = No. m-moles HB consumed

$$= (40 \text{ ml}) (0.10\, M) = 4.0 \text{ m-moles}$$

The hydrolysis of B^- gives $B^- + H_2O \rightleftharpoons HB + OH^-$. $K_h = K_w/K_i = 1.02 \times 10^{-10}$. Let x equal the number of moles of B^- hydrolyzing. $x = [HB] = [OH^-]$; 4.0 m-moles/80 ml $- x = [B^-]$.

$$1.02 \times 10^{-10} = \frac{(x)\,(x)}{0.05 - x}$$

$$x = 2.26 \times 10^{-6} = [OH^-]$$

$$pH = 8.35$$

(e) The amount of OH^- resulting from the hydrolysis of B^- is negligibly small compared with the OH^- from 1.0 ml of 0.10 M

KOH: 1.0 ml KOH contains 0.10 m-mole. From part d, we know that the hydrolysis produces only 2.26×10^{-6} mole OH^-/liter. The concentration of OH^- from KOH will be: $[OH^-] = 0.10$ m-mole/81 ml $= 0.0012$, which is much larger than 2.26×10^{-6}. Therefore, the hydrolysis of B^- can be ignored and the pH calculated from the KOH contribution.

$$pOH = -\log 0.0012 = 2.91 \qquad pH = 11.09$$

78. The acids $(CH_3)_2 AsO_2 H$, $ClCH_2 COOH$, and $CH_3 COOH$ have ionization constants 6.40×10^{-7}, 1.40×10^{-3}, and 1.76×10^{-5}, respectively. Which acid would be best for preparing a buffer solution at $pH = 6.50$? How many grams of this acid and of NaOH would be needed to prepare 1.00 liter of buffer solution in which the sum of the concentrations of the acid and its conjugate base is equal to $1.00\ M$?

Soln: To be most efficient, a buffer solution should contain a concentration ratio of ion to acid or ion to base as nearly equal to 1.0 as possible. For acids in general,

$$K_i = \frac{[H^+]\ [A^-]}{[HA]}$$

$K_i = [H^+]$ when $[A^-] = [HA]$. The $[H^+]$ for a solution with $pH = 6.5$ is 3.16×10^{-7}. Therefore, $(CH_3)_2 AsO_2 H$ is most appropriate. The ratio of $(CH_3)_2 AsO_2^-$ to $(CH_3)AsO_2 H$ is

$$K_i = 6.4 \times 10^{-7} = (3.16 \times 10^{-7}) \frac{[(CH_3)_2 AsO_2^-]}{[(CH_3)_2 AsO_2 H]}$$

$$[(CH_3)_2 AsO_2^-] = 2.02[(CH_3)_2 AsO_2 H]$$

The sum of the amounts of ion and acid to be used is 1.0 mole. Therefore, 2.02 times as much ion is needed as acid: The solution must contain 1.0 mole/3.02 = 0.331 mole of acid and 0.669 mole of ion. The solution can be prepared by allowing one mole of the acid (138 g) to react with 0.669 mole of NaOH.

Wt NaOH required = (0.669) (40.0 g/mole) = 26.8 g.

79. Calculate the concentration of each of the ions in solution of $0.400\ M$ $H_2 SO_4$. (First step is essentially 100% dissociated into ions; $K_2 = 1.2 \times 10^{-2}$.)

Soln: $H_2 SO_4$ ionizes in two steps:

$$H_2 SO_4 \rightarrow H^+ + HSO_4^-; \quad K_1 \text{ is very large}$$

$$HSO_4^- \rightleftharpoons H^+ + SO_4^{2-}; \quad K_2 = 1.2 \times 10^{-2}$$

It is assumed that the first ionization is complete. However, the total $[H^+]$ is the sum of the $[H^+]$ from the first and second steps.

Since HSO_4^- is not completely ionized, let x equal the number of moles of HSO_4^- that ionize. Then $[HSO_4^-] = 0.400 - x$; $[H^+] = 0.400 + x$; $[SO_4^{2-}] = x$. Substitution into K_e yields

$$\frac{(0.400 + x)\,(x)}{(0.400 - x)} = 1.2 \times 10^{-2}$$

Is x small relative to 0.400? No.

Multiply through both sides of the equation, transpose terms, and solve for x via the quadratic formula:

$$x^2 + 0.412x - 0.0048 = 0$$

Solving the quadratic equation gives the roots: $x = -0.847/2$ and $0.023/2$. Since x cannot be negative, the value for x is 0.012.

$$[H^+] = 0.400 + 0.012 = 0.41$$

$$[SO_4^{2-}] = 0.011$$

$$[HSO_4^-] = 0.400 - 0.012 = 0.39$$

The $[OH^-]$ is calculated from K_w.

$$K_w = [H^+]\,[OH^-] = 1 \times 10^{-14}$$

$$(0.41)\,[OH^-] = 1 \times 10^{-14}$$

$$[OH^-] = 2.4 \times 10^{-14}$$

UNIT XI

The Solubility Product Principle

19

INTRODUCTION

Several factors influence the equilibrium of ionic substances in water. The substances examined here dissolve to produce saturated aqueous solutions containing ions from the compound in dynamic equilibrium with the undissolved solute. Understanding the solubility product concept involves some complicating features, such as the effect of two weak electrolytes that simultaneously compete for ions. Fractional precipitation, dissolution of precipitates by various means, and calculations involving complex ions are all treated. Knowledge of the last-mentioned items is particularly important in chemical analysis.

FORMULAS AND DEFINITIONS

Saturated solution   Solution that is in equilibrium with undissolved solute.

K_{sp} An equilibrium constant based on the molar concentrations of ions in a saturated solution containing a slightly soluble electrolyte. Each ion concentration must be raised to the power corresponding to its stoichiometric coefficient as shown in the equilibrium expression.

Salt effect   A condition in which the solubility of a slightly soluble substance is increased by the presence of high concentrations of an added electrolyte.

K_d The dissociation constant is an equilibrium constant that is a measure of the stability of a complex ion in solution.

PROBLEMS

1. Calculate the solubility product constant of each of the following from the solubility given: (a) $BaCO_3$ (9.0×10^{-5} mole/liter), (c) Ag_2SO_4 (4.47 g/liter)

   Soln: (a) The compound dissolves and dissociates: $\underline{BaCO_3} \rightleftharpoons Ba^{2+} + CO_3^{2-}$. Dissolution produces one mole each of the ions Ba^{2+} and CO_3^{2-}. If the solubility is 9.0×10^{-5} mole/liter, concentration of the ions Ba^{2+} and CO_3^{2-} equals 9.0×10^{-5} mole/liter:

   $$K_{sp} = [Ba^{2+}] \, [CO_3^{2-}]$$
   $$= [9.0 \times 10^{-5}] \, [9.0 \times 10^{-5}] = 8.1 \times 10^{-9}$$

   (c) Dissolution of Ag_2SO_4 occurs as $\underline{Ag_2SO_4} \rightleftharpoons 2Ag^+ + SO_4^{2-}$. GFW of $Ag_2SO_4 = 312$ g; solubility $= \overline{4.47 \text{ g/liter}} \times 1\text{mole}/312 \text{ g} = 1.43 \times 10^{-2}$ mole/liter.

   Two moles of Ag^+ and one mole of SO_4^{2-} are produced for each mole of Ag_2SO_4 that dissolves. Hence, $[Ag^+] = 2(1.43 \times 10^{-2}$ mole/liter), and $[SO_4^{2-}] = 1.43 \times 10^{-2}$ mole/liter. The expression for K_{sp} is:

   $$K_{sp} = [Ag^+]^2 \, [SO_4^{2-}]$$
   $$= [2(1.43 \times 10^{-2})]^2 [1.43 \times 10^{-2}]$$
   $$= 1.17 \times 10^{-5}$$

2. Calculate the molar solubility for each of the following substances from its solubility product constant:
   (a) $SrCO_3$; $K_{sp} = 9.42 \times 10^{-10}$.
   (c) Ag_3PO_4; $K_{sp} = 1.8 \times 10^{-18}$.

   Soln: (a) Dissolving one formula unit of $SrCO_3$ produces one each of the ions Sr^{2+} and CO_3^{2-}. $\underline{SrCO_3} \rightleftharpoons Sr^{2+} + CO_3^{2-}$; $K_{sp} = [Sr^{2+}] [CO_3^{2-}]$.

   Let x equal the solubility of $SrCO_3$ in moles per liter.

   $$x = [Sr^{2+}] = [CO_3^{2-}]$$
   $$K_{sp} = (x)(x) = 9.42 \times 10^{-10}$$
   $$x^2 = 9.42 \times 10^{-10}$$
   $$x = 3.07 \times 10^{-5} \text{ mole/liter}$$

(c) $Ag_3PO_4 \rightleftharpoons 3Ag^+ + PO_4{}^{3-}$; $K_{sp} = [Ag^+]^3 [PO_4{}^{3-}]$.

Let x equal the molar solubility of Ag_3PO_4. Since dissolving Ag_3PO_4 produces three Ag^+ and one $PO_4{}^{3-}$, $[Ag^+] = 3x$ and the $[PO_4{}^{3-}] = x$.

$$K_{sp} = (3x)^3 (x) = 1.8 \times 10^{-18}$$
$$27x^4 = 1.8 \times 10^{-18}$$
$$x^4 = 6.67 \times 10^{-20}$$

The fourth root of a number is equivalent to the square root of the square root of the number. In this case, $x = 1.6 \times 10^{-5}$ mole/liter.

4. Calculate the $[Mg^{2+}]$ required to start the precipitation of MgF_2 from a solution that is 0.0040 M in fluoride ions ($K_{sp} = 6.4 \times 10^{-9}$).

Soln: Precipitation of MgF_2 will occur when the product of $[Mg^{2+}]$ $[F^-]^2$ is greater than K_{sp} of MgF_2. In a solution 0.0040 M in fluoride ions,

$$K_{sp} = [Mg^{2+}] [F^-]^2 = 6.4 \times 10^{-9}$$

the $[Mg^{2+}]$ required to initiate precipitation must be greater than

$$[Mg^{2+}] = \frac{6.4 \times 10^{-9}}{[F^-]^2} = \frac{6.4 \times 10^{-9}}{(0.004)^2} = 4.0 \times 10^{-4} \, M$$

6. The K_{sp} of HgS is 3×10^{-53}. Calculate the maximum concentration of sulfide ion that can exist in a solution that is 1.0×10^{-10} M in $Hg(NO_3)_2$.

Soln: It is assumed that compounds like $Hg(NO_3)_2$ are completely dissociated in aqueous solution. Therefore, $[Hg^{2+}]$ in this solution is 1×10^{-10} M. According to the K_{sp} of HgS, the concentration of the ion S^{2-} is found as follows:

$$HgS \rightleftharpoons Hg^{2+} + S^{2-}$$
$$K_{sp} = [Hg^{2+}] [S^{2-}] = 3 \times 10^{-53}$$
$$[S^{2-}] = \frac{3 \times 10^{-53}}{[Hg^{2+}]}$$
$$= \frac{3 \times 10^{-53}}{1 \times 10^{-10}} = 3 \times 10^{-43} \, M$$

8. Calculate the solubility of AgCl in mg/100 ml. ($K_{sp} = 1.8 \times 10^{-10}$.)

Soln: K_{sp} constants are calculated from molar solubility values. Therefore, the solubility of a compound in terms of mass must be calculated from its molar solubility.

$$AgCl \rightleftharpoons Ag^+ + Cl^-$$
$$K_{sp} = [Ag^+] [Cl^-] = 1.8 \times 10^{-10}$$

Let x equal the molar solubility of AgCl.

$$[Ag^+] = [Cl^-] = x$$

$$K_{sp} = (x)(x) = x^2 = 1.8 \times 10^{-10}$$

$$x = 1.3 \times 10^{-5}$$

GFW AgCl = 143.5 g;

Wt AgCl/liter = 1.3×10^{-5} mole/liter $\times$ 143.5 g/mole

$$= 1.9 \times 10^{-3} \text{ g/liter}$$

Wt (mg/100 ml) = 1.9×10^{-3} g/liter $\times 10^3$ mg/g $\times$ 0.1 liter

$$= 0.19 \text{ mg/100 ml}$$

9. A precipitate of AgCl was washed with 100 ml of 0.10 M HCl solution and then with 100 ml of distilled water. Calculate the amount of AgCl which dissolved with each washing, assuming that the wash liquid became saturated with AgCl. (K_{sp} of AgCl = 1.8×10^{-10}.)

Soln: AgCl $\rightleftharpoons$ Ag$^+$ + Cl$^-$; K_{sp} = 1.8×10^{-10} = $[Ag^+][Cl^-]$. Calculate the amount of AgCl that will dissolve in 100 ml of each solution according to the K_{sp} expression. In 0.10 M HCl,

$$[Ag^+] = \frac{1.8 \times 10^{-10}}{[Cl^-]}$$

$$= \frac{1.8 \times 10^{-10}}{0.10 \ M} = 1.8 \times 10^{-9} \text{ mole/liter}$$

Amount of Ag$^+$ in 100 ml = 1.8×10^{-10} mole.

For H$_2$O, let x equal the molar solubility of AgCl.

$$x = [Ag^+] = [Cl^-] \qquad \text{and} \qquad \frac{x}{10} = \frac{\text{solubility}}{100 \text{ ml}}$$

$$K_{sp} = (x)(x) = x^2 = 1.8 \times 10^{-10}$$

$$x = 1.3 \times 10^{-5}$$

$$\frac{x}{10} = 1.3 \times 10^{-6} \text{ mole}$$

12. A 300-ml volume of 2×10^{-7} M Hg$_2$(NO$_3$)$_2$ solution and 300 ml of 3×10^{-5} M KI solution are mixed. Will Hg$_2$I$_2$ precipitate? Explain your answer. (K_{sp} of Hg$_2$I$_2$ is 4.5×10^{-29}.)

Soln: The species in which mercury appears to have a positive one oxidation state is always written Hg$_2$$^{2+}$; that is, it is found as a unit consisting of two mercury ions with a positive two charge overall. Precipitation will occur if [Hg$_2$$^{2+}$][I$^-$]^2 is greater than K_{sp}. Calculate the ion concentration that exists immediately after mixing.

$$[Hg_2{}^{2+}] = \frac{(300\ ml)\ (2 \times 10^{-7}\ M)}{600\ ml} = 1 \times 10^{-7}\ M$$

$$[I^-] = \frac{(300\ ml)\ (3 \times 10^{-5}\ M)}{600\ ml} = 1.5 \times 10^{-5}\ M$$

$$K_{spHg_2I_2} = [Hg_2{}^{2+}]\ [I^-]^2 = 4.5 \times 10^{-29}$$

Is the ion product greater than K_{sp}?

$$[1 \times 10^{-7}\ M]\ [1.5 \times 10^{-5}\ M]^2 = 2.3 \times 10^{-17}$$

$$2.3 \times 10^{-17} \gg 4.5 \times 10^{-29}$$

Therefore, precipitation will occur.

16. A solution which is $0.10\ M$ in both NaI and $Na_2\ SO_4$ is treated with solid $Pb(NO_3)_2$. Which compound, PbI_2 or $PbSO_4$, will precipitate first? What is the concentration of the anion of the least soluble compound when the more soluble one starts to precipitate? (K_{sp} of $PbI_2 = 8.7 \times 10^{-9}$; K_{sp} of $PbSO_4 = 1.8 \times 10^{-8}$.)

Soln: Calculate the molar solubility of each compound: the least soluble compound will precipitate first. Let x equal the molar solubility of PbI_2: $\underline{PbI_2 \rightleftharpoons Pb^{2+} + 2I^-}$; $K_{sp} = 8.7 \times 10^{-9}$.

$$x = [Pb^{2+}]; \qquad 2x = [I^-]$$

$$K_{sp} = (x)\ (2x)^2 = 8.7 \times 10^{-9}$$

$$4x^3 = 8.7 \times 10^{-9}$$

$$x^3 = 2.18 \times 10^{-9}$$

The cube root of 2.18×10^{-9} can be extracted by using a slide rule or logarithms.

$$x = 1.3 \times 10^{-3}\ mole/liter$$

For $PbSO_4$, let x equal its molar solubility. $\underline{PbSO_4 \rightleftharpoons [Pb^{2+}]}$ $[SO_4{}^{2-}]; K_{sp} = 1.8 \times 10^{-8}$.

$$x = [Pb^{2+}] = [SO_4{}^{2-}]$$

$$x^2 = K_{sp} = 1.8 \times 10^{-8}$$

$$x = 1.34 \times 10^{-4}\ mole/liter$$

Comparing the molar solubilities of the compounds, we see that $PbSO_4$ is less soluble than PbI_2 and will precipitate first. In fact, $PbSO_4$ will continue to precipitate until the ion product, $[Pb^{2+}]$ $[I^-]^2$, exceeds K_{sp} of PbI_2. Since $[I^-] = 0.10\ M$, $[Pb^{2+}]$ without PbI_2 precipitation can be no greater than

$$[Pb^{2+}] = \frac{8.7 \times 10^{-9}}{(0.10)^2} = 8.7 \times 10^{-7}\ M$$

Therefore, $[SO_4^{2-}]$ will be diminished to

$$[SO_4^{2-}] = \frac{1.8 \times 10^{-8}}{8.7 \times 10^{-7}} = 0.021\ M$$

before PbI_2 will precipitate.

19. A 50-ml volume of solution containing 0.95 g of $MgCl_2$ is mixed with an equal volume of 1.8 M aqueous ammonia. What weight of solid $NH_4\,Cl$ must be added to the resulting solution to prevent precipitation of $Mg(OH)_2$? (K_{sp} of $Mg(OH)_2$ = 1.5×10^{-11}; K_i of NH_3 is 1.8×10^{-5}.)

   Soln: Preventing precipitation of $Mg(OH)_2$ in an aqueous ammonia solution requires simultaneous consideration of the ammonia equilibrium and K_{sp} of $Mg(OH)_2$. First, calculate the maximum $[OH^-]$ that sustains the $Mg(OH)_2$ equilibrium. $\underline{Mg(OH)_2 \rightleftharpoons Mg^{2+} + 2OH^-}$; $K_{sp} = 1.5 \times 10^{-11}$.

$$K_{sp} = [Mg^{2+}]\ [OH^-]^2 = 1.5 \times 10^{-11}$$

   GFW of $MgCl_2$ = 95.0 g; no. moles $MgCl_2$ = 0.95 g per 95.0 g/mole = 0.010 mole.

$$\text{Initial }[Mg^{2+}] = \frac{0.010\text{ mole}}{100\text{ ml}} = 0.10\ M$$

$$[NH_3] = \frac{(1.8\ M)\ (50\text{ ml})}{100\text{ ml}} = 0.90\ M$$

   The maximum $[OH^-]$ in a 0.10 M Mg^{2+} solution is determined from K_{sp} of $Mg(OH)_2$:

$$[OH^-]^2 = \frac{1.5 \times 10^{-11}}{0.10} = 1.5 \times 10^{-10}$$

$$[OH^-] = 1.22 \times 10^{-5}\ M$$

   From the ammonia equilibrium, calculate the $[NH_4^+]$ required to keep the $[OH^-]$ at 1.22×10^{-5} M: $NH_3 + H_2O \rightleftharpoons NH_4^+ + OH^-$.

$$K_i = 1.8 \times 10^{-5} = \frac{[NH_4^+]\ [OH^-]}{[NH_3]}$$

$$= \frac{[NH_4^+]\ (1.22 \times 10^{-5})}{(0.90)}$$

$$[NH_4^+] = 1.32\text{ moles/liter}$$

   Wt of $NH_4\,Cl/100$ ml = 1.32 moles/liter $\times$ 0.10 liter $\times$ 53.5 g/mole

$$= 7.1\text{ g}$$

21. What pH is necessary for initiating the precipitation of lead(II) hydroxide from a solution of lead(II) nitrate which contains 0.662 mg of lead(II) ion per ml? (K_{sp} of $Pb(OH)_2$ = 2.8×10^{-16}.)

Soln: $\underline{Pb(OH)_2} \rightleftharpoons Pb^{2+} + 2OH^-$; $K_{sp} = 2.8 \times 10^{-16}$.

$$[Pb^{2+}] = \frac{\text{m-moles}}{\text{ml}} = \frac{0.662 \text{ mg}}{\text{ml}} \times \frac{1.0 \text{ m-mole}}{207.2 \text{ mg}} = 3.2 \times 10^{-3} \ M$$

$$[OH^-]^2 = \frac{2.8 \times 10^{-16}}{3.2 \times 10^{-3}} = 8.76 \times 10^{-14}$$

$$[OH^-] = 2.96 \times 10^{-7} \ M$$

$$pOH = 6.53$$

$$pH = 7.47$$

23. What is the maximum pH that a $0.075 \ M$ $Fe(NO_3)_2$ solution can have and not have FeS precipitate when the solution is saturated with $H_2 S$? A saturated $H_2 S$ solution is $0.10 \ M$ in $H_2 S$. (K_{sp} of FeS $= 1 \times 10^{-19}$; K_1 and K_2 for $H_2 S$ are 1.0×10^{-7} and 1.3×10^{-13}, respectively.)

   Soln: The FeS and $H_2 S$ equilibria must be considered simultaneously. The maximum sulfide concentration that can be tolerated is
   $\underline{FeS} \rightleftharpoons Fe^{2+} + S^{2-}$; $K_{sp} = 1 \times 10^{-19}$.

   $$[S^{2-}] = \frac{1 \times 10^{-19}}{0.075} = 1.33 \times 10^{-18}$$

   Since the sulfide must come from the ionization of $H_2 S$, the following equilibrium prevails: $H_2 S \rightleftharpoons 2H^+ + S^{2-}$.

   $$K = K_1 K_2 = \frac{[H^+]^2 [S^{2-}]}{[H_2 S]}$$

   $$= 1.0 \times 10^{-7} \times 1.3 \times 10^{-13}$$

   $$= 1.3 \times 10^{-20}$$

   $$[H^+]^2 = \frac{1.3 \times 10^{-20} \ [H_2 S]}{[S^{2-}]}$$

   $$= \frac{(1.3 \times 10^{-20})(0.10)}{(1.33 \times 10^{-18})}$$

   $$= 9.8 \times 10^{-4}$$

   $$[H^+] = 3.13 \times 10^{-2}$$

   $$pH = 1.5$$

24. A solution which is $0.10 \ M$ in Cd^{2+} and $0.30 \ M$ in HCl is saturated with $H_2 S$. What concentration of Cd^{2+} will remain in solution? (K_{sp} of CdS $= 3.6 \times 10^{-29}$. Do not neglect the hydrogen ions produced by the reaction of Cd^{2+} with $H_2 S$.)

   Soln: The CdS and $H_2 S$ equilibria must be satisfied simultaneously. In this case, the amount of CdS left in solution is extremely small compared with the initial 0.10 M Cd^{2+} solution. We shall assume, therefore, that the reaction of Cd^{2+} with H_2S goes to completion, and treat the solution in terms of the maximum $[S^{2-}]$ that can exist in the solution with a specified $[H^+]$. According to the following reaction, 0.10 mole of Cd^{2+} will react with H_2S to release 0.20 mole of H^+: $Cd^{2+} + H_2S \rightarrow CdS + 2H^+$. The hydrogen ion concentration from the $HCl(0.30\,M)$ plus $[H^+]$ from the reaction is $0.50\,M$. The $[S^{2-}]$ can be calculated from the H_2S equilibrium:

$$\frac{[H^+]^2\,[S^{2-}]}{[H_2S]} = 1.3 \times 10^{-20} = K \text{ (from Problem 23)}$$

$$[S^{2-}] = \frac{(1.3 \times 10^{-20})\,[H_2S]}{[H^+]^2}$$

$$= \frac{(1.3 \times 10^{-20})\,(0.10)}{(0.50)^2}$$

$$= 5.2 \times 10^{-21}$$

The Cd^{2+} ion concentration in solution is

$$K_{sp} = [Cd^{2+}]\,[S^{2-}] = 3.6 \times 10^{-29}$$

$$[Cd^{2+}] = \frac{3.6 \times 10^{-29}}{5.2 \times 10^{-21}} = 6.9 \times 10^{-9}\,M$$

28. The K_{sp} of $Mn(OH)_2$ is 4.5×10^{-14}. What is the pH of a saturated $Mn(OH)_2$ solution?

Soln: $Mn(OH)_2 \rightleftharpoons Mn^{2+} + 2OH^-$; $K_{sp} = 4.5 \times 10^{-14}$.
Let x equal the molar solubility of $Mn(OH)_2$:

$$[OH^-] = 2x; \qquad [Mn^{2+}] = x$$

$$(x)\,(2x)^2 = 4.5 \times 10^{-14}$$

$$4x^3 = 4.5 \times 10^{-14}$$

$$x^3 = 1.13 \times 10^{-14}$$

$$x = 2.24 \times 10^{-5}$$

$$[OH^-] = 2x = 4.48 \times 10^{-5}$$

$$p\text{OH} = 4.35$$

$$p\text{H} = 9.65$$

31. What is the molar solubility of MgF_2 in a $0.30\,M$ HCl solution? (K_{sp} of $MgF_2 = 6.4 \times 10^{-9}$; K_i of HF $= 7.2 \times 10^{-4}$.)

Soln: K_{sp} for MgF_2 is 6.4×10^{-9}. This small value indicates that the molar solubility of MgF_2 is quite small. The compound, however, dissociates in a solution in which the hydrogen ion concentration is approximately 0.30 M. As a consequence, most of the fluoride ions produced associate with hydrogen ions to produce HF, according to the reaction $H^+ + F^- \rightleftharpoons HF$.

This equilibrium is the reverse of the ionization of HF, which has a K_i of 7.2×10^{-4}:

$$HF \rightleftharpoons H^+ + F^-; \quad K_i = 7.2 \times 10^{-4}$$

At equilibrium the following relations exist:

$$\underline{MgF_2} \rightleftharpoons Mg^{2+} + 2F^-; \quad K_{sp} = 6.4 \times 10^{-9}$$

$$[Mg^{2+}] \, [F^-]^2 = 6.4 \times 10^{-9}$$

$$HF \rightleftharpoons H^+ + F^-; \quad K_i = 7.2 \times 10^{-4}$$

Dissociation of one mole of MgF_2 produces two moles of F^-. The total amount of fluorine in solution equals the sum of $[F^-]$ and $[HF]$:

$$[F]_{total} = [F^-] + [HF]$$

$$[H^+] = 0.30 \, M - [HF]$$

$$[Mg^{2+}] = \frac{[F]_{total}}{2} = \frac{[F^-] + [HF]}{2}$$

The molar solubility of MgF_2 is calculated by the simultaneous solution of the HF and MgF_2 equilibria expressions for $[Mg^{2+}]$. Solve the HF equilibrium expression for $[HF]$ in terms of F^-.

$$\frac{[H^+] \, [F^-]}{[HF]} = 7.2 \times 10^{-4} = \frac{[0.30 - [HF]] \, [F^-]}{[HF]}$$

$$[HF] = \frac{0.30 \, [F^-]}{7.2 \times 10^{-4} + [F^-]}$$

Substitute this value for $[HF]$ in the K_{sp} expression for MgF_2 as:

$$\left[\frac{[F^-] + [HF]}{2} \right] [F^-]^2 = 6.4 \times 10^{-9}$$

$$\left[\frac{[F^-] + (0.30[F^-]/(7.2 \times 10^{-4} + [F^-]))}{2} \right] [F^-]^2 = 6.4 \times 10^{-9}$$

Rearrangement of this equation yields

$$[F^-]^4 + 0.30[F^-]^3 - 1.28 \times 10^{-8} [F^-] - 9.2 \times 10^{-12} = 0$$

An equation such as this usually cannot be factored or otherwise solved by simple algebraic techniques; an approximation method must be used for solution. A calculator is very helpful for solutions of this type. The best value for $[F^-]$ is the value that makes the expression most nearly equal to zero. Unfortunately there is no straightforward method that works best for approximation calculations. A sensible approach is to approximate the value of $[F^-]$ by solving the K_{sp} expression for $[F^-]$, ignoring the HCl–HF component. Realizing that more MgF_2 will dissolve in the HCl solution than in water alone, you know that $[F^-]$ will be less in HCl than in water. The calculation indicates that $[F^-]$ is between 10^{-4} M and 10^{-3} M. After a few successive substitutions of values into the fourth-order equation, you should arrive at a value for $[F^-]$ that is nearly equal to 3.6×10^{-4}. Some representative substitutions are shown in the accompanying chart.

$[F^-]$	Value
1×10^{-4}	-1.02×10^{-11}
2×10^{-4}	$-2.6 \ \times 10^{-10}$
3.6×10^{-4}	$2.2 \ \times 10^{-13}$
5×10^{-4}	$2.2 \ \times 10^{-11}$
8×10^{-4}	1.35×10^{-10}
1×10^{-3}	$2.8 \ \times 10^{-10}$

Substituting $[F^-] = 3.6 \times 10^{-4}$ in the K_{sp} expression and solving for $[Mg^{2+}]$ gives

$$[Mg^{2+}] = \frac{6.4 \times 10^{-9}}{(3.6 \times 10^{-4})^2} = 4.9 \times 10^{-2} = 0.05 \ M$$

34. A solution is $0.10 \ M$ in $ZnCl_2$. What concentration of hydrogen ion must be present so that no zinc(II) sulfide will precipitate when the solution is saturated with hydrogen sulfide? A saturated solution of H_2S is $0.10 \ M$.

Soln: The sulfide ion concentration that originates from H_2S depends on the zinc ion concentration, which is $0.10 \ M$ in Zn^{2+}.

$$ZnS \rightleftharpoons Zn^{2+} + S^{2-}$$

$$K_{sp} = [Zn^{2+}] \ [S^{2-}] = 1.1 \times 10^{-21}$$

$$[S^{2-}] = \frac{1.1 \times 10^{-21}}{0.10} = 1.1 \times 10^{-20}$$

$$H_2S \rightleftharpoons 2H^+ + S^{2-}; \quad K_i = 1.3 \times 10^{-20}$$

$$[H^+]^2 = \frac{K_i[H_2S]}{[S^{2-}]}$$

$$= \frac{(1.3 \times 10^{-20})(0.10)}{1.1 \times 10^{-20}}$$

$$= 1.18 \times 10^{-1}$$

$$[H^+] = 0.34 \ M$$

36. It is found that 1.62×10^{-9} mole of PbS will dissolve in a liter of a solution which has a pH of 1.50 and which contains H_2S at $0.040\ M$ concentration. Using the dissociation constants for hydrogen sulfide, calculate the K_{sp} for lead sulfide. Compare the calculated value with the value given in Appendix E.

Soln: The sulfide ion concentration depends on the H_2S equilibrium at the given acidity conditions.

$$\underline{PbS} \rightleftharpoons Pb^{2+} + S^{2-}; \qquad K_{sp} = [Pb^{2+}]\ [S^{2-}]$$

$$H_2S \rightleftharpoons 2H^+ + S^{2-}; \qquad K_i = 1.3 \times 10^{-20}$$

$$pH = 1.5; \qquad [H^+] = 3.16 \times 10^{-2}$$

$$[S^{2-}] = \frac{(1.3 \times 10^{-20})[H_2S]}{[H^+]^2}$$

$$= \frac{(1.3 \times 10^{-20})\ (0.04)}{(3.16 \times 10^{-2})^2}$$

$$= 5.2 \times 10^{-19}$$

The Pb^{2+} ion concentration is $1.62 \times 10^{-9}\ M$ and the K_{sp} of PbS equals $[Pb^{2+}]\ [S^{2-}]$.

$$K_{sp} = (1.62 \times 10^{-9})\ (5.2 \times 10^{-19}) = 8.4 \times 10^{-28}$$

38. Calculate the concentration of Ni^{2+} in a $1.0\ M$ solution of $[Ni(NH_3)_6](NO_3)_2$. (K_d for $[Ni(NH_3)_6]^{2+} = 5.7 \times 10^{-9}$.)

Soln: The compound dissolves in water, giving

$$[Ni(NH_3)_6](NO_3)_2 \rightarrow Ni(NH_3)_6{}^{2+} + 2NO_3{}^-$$

Complex ions, such as $Ni(NH_3)_6{}^{2+}$, undergo dissolution and reach equilibrium in water. In this case, the dissolution is given by:

$$[Ni(NH_3)_6]^{2+} \rightleftharpoons Ni^{2+} + 6NH_3; \qquad K_d = 5.7 \times 10^{-9}$$

$$K_d = \frac{[Ni^{2+}]\ [NH_3]^6}{[Ni(NH)_3{}^{2+}]} = 5.7 \times 10^{-9}$$

Let x equal the number of moles of $[Ni(NH_3)_6]^{2+}$ dissociating.

$$x = [Ni^{2+}]; 6x = [NH_3]; \qquad 1.0 - x = [Ni(NH_3)_6{}^{2+}]$$

$$\frac{(x)\ (6x)^6}{(1.0 - x)} = 5.7 \times 10^{-9}$$

Assume x to be small compared with $1.0\ M$ and simplify the expression to

$$6^6 x^7 = 5.7 \times 10^{-9}$$

$$x^7 = \frac{5.7 \times 10^{-9}}{46,700} = 1.22 \times 10^{-13}$$

Extracting the seventh root of a number requires using logarithms, as indicated below. The process is simplified by rewriting the number in scientific notation to include an exponent on ten that is evenly divisible by 7.

$$x^7 = 12.2 \times 10^{-14}$$

$$x = (x^7)^{1/7} = (12.2)^{1/7} \times (10^{-14})^{1/7}$$

$$= (12.2)^{1/7} \times 10^{-2}$$

$$\log(12.2)^{1/7} = \tfrac{1}{7} \log 12.2$$

$$= \tfrac{1}{7}(1.0864) = 0.1552$$

$$(12.2)^{1/7} = 10^{0.1552} = 1.43$$

$$\therefore x = 0.014 \, M$$

40. Calculate the minimum number of moles of cyanide ion that must be added to 100 ml of solution to dissolve 2×10^{-2} mole of AgCN. (K_{sp} for AgCN is 1.2×10^{-16}; K_d for $[Ag(CN)_2]^-$ is 1×10^{-20}.)

Soln: Initial AgCN concentration = 0.02 mole/0.10 liter − 0.20 M. The $Ag(CN_2)^-$ and AgCN equilibria must be satisfied simultaneously. Since the ion $Ag(CN)_2^-$ is very stable ($K_d = 1 \times 10^{-20}$), essentially all the AgCN will be converted to the complex. The amount of CN^- to be added must equal the amount required to form the $Ag(CN)_2^-$ plus the amount required to maintain the equilibrium. The $[Ag^+]$ in solution applies to both equilibria when both solid AgCN and $Ag(CN)_2^-$ are present in solution.

$$K_{sp} = [Ag^+][CN^-] = 1.2 \times 10^{-16}$$

$$[Ag^+] = \frac{1.2 \times 10^{-16}}{[CN^-]} = \frac{1.2 \times 10^{-16}}{0.20} = 6.0 \times 10^{-16}$$

$$K_d = \frac{[Ag^+][CN^-]^2}{[Ag(CN)_2^-]} = 1.0 \times 10^{-20}$$

$$[Ag(CN)_2^-] = 0.20 \, M$$

$$[CN^-]^2 = \frac{(1.0 \times 10^{-20})(0.200)}{6 \times 10^{-16}}$$

$$[CN^-] = 1.8 \times 10^{-3} \, M$$

Amount to be added = (0.20 + 0.0018) mole/liter = 0.20 mole/liter. To 100 ml, add 2×10^{-2} mole.

43. In a titration of cyanide ion, 28.72 ml of 0.0100 M AgNO₃ is added before precipitation begins. How many grams of NaCN were in the original sample? What is the insoluble material? What is the nature of the silver species before precipitation begins?

Soln: The reaction of Ag^+ with CN^- goes to completion producing the $Ag(CN)_2^-$ complex. Adding excess Ag^+ to the complex produces AgCN, which precipitates. The amount of CN^- initially in solution is determined from the reaction, $Ag^+ + 2CN^- \rightleftharpoons Ag(CN)_2^-$.

Ag^+ used = (0.02872 liter) (0.0100 M) = 2.872×10^{-4} mole

Moles CN^- required = (2) $(2.872 \times 10^{-4}$ mole) = 5.74×10^{-4}

Wt NaCN = $(5.74 \times 10^{-4}$ mole) (49.0 g/mole) = 0.0281 g

UNIT XII

Chemical Thermodynamics

20

INTRODUCTION

The subject matter of chemical thermodynamics involves three basic questions in which chemists are interested: (1) Will two or more substances react? (2) If a reaction does occur, what is the associated energy change? (3) If the reaction occurs, what will be the equilibrium concentrations of the reactants and products? To develop thermodynamics and aid us in answering the above questions, we use two fundamental laws of nature: (1) Systems tend to a state of minimum potential energy. (2) Systems tend to a state of maximum disorder. The three laws of thermodynamics incorporate these ideas into a form from which we can develop information useful for answering the opening three questions.

FORMULAS AND DEFINITIONS

Chemical thermodynamics That branch of chemistry that studies the energy transformations and transfers that accompany chemical and physical changes.

System That part of the universe we are studying and with whose properties we are concerned.

Surroundings All of the universe except the system we are studying.

State The condition of the system, defined by n, P, V, and T.

First law of thermodynamics A statement of the law of conservation of energy; the total amount of energy in the universe is constant. Mathematically, $E_{sys} = E_2 - E_1 = q - w$. Here E_{sys} is the internal energy change of the system due to a change in state, q is heat, and w is work.

Heat (q) A positive sign indicates a heat increase in the system; a negative sign corresponds to loss of heat from the system.

Work (w) A positive sign indicates work done by the system; a negative sign corresponds to work done on the system. Work has a pressure-volume equivalent defined as $w = P(V_2 - V_1)$, where P is the pressure restraining the system, V_1 is the initial volume, and V_2 is the final volume.

Enthalpy (H) The heat content or enthalpy of the system. The change in enthalpy ΔH is the quantity of heat absorbed when a reaction takes place at constant pressure; therefore, $\Delta H = q$. By definition, $\Delta H = \Delta E + \Delta(PV)$, or $\Delta H = \Delta E + P\,\Delta V$ for a constant pressure process.

Standard state Specific set of conditions agreed to for facilitating handling of data. The standard state of a pure substance is taken as $25°C$ ($298.15°K$) and one atmosphere pressure.

$\Delta H°_{f298}$ Standard molar enthalpy of formation. The change in enthalpy when one mole of a pure substance is formed from the free elements in their most stable state under standard conditions. For any free element in its most stable form, the value of the standard molar enthalpy is zero.

Hess's law For any process that can be considered the sum of several stepwise processes, the enthalpy change for the total process must equal the sum of the enthalpy changes for the various steps.

Bond energy A measure of the heat of formation of a compound, determined by summing the strengths of the individual chemical bonds formed in a reaction and subtracting the strengths of the bonds broken in the reaction.

Entropy (S) A measure of the randomness of a system. The importance of entropy lies in our ability to predict the direction of a chemical process if both the entropy of the system and the entropy of the surroundings are known.

Second law of thermodynamics In any spontaneous change, the entropy of the universe increases.

Third law of thermodynamics The entropy of any pure perfect crystalline substance at the absolute zero of temperature ($0°K$) equals zero. Basically, this law allows establishing a beginning point, or zero point, for entropy measurements.

Gibbs free energy change (ΔG) Perhaps the most useful function of thermodynamics. It is the maximum amount of useful work that can be accomplished by a reaction at constant temperature and pressure. This quantity can also be used to determine the direction of a chemical process using only information about the system. This occurs since reactions tend to proceed to a state of maximum disorder (positive ΔS) and minimum energy (negative ΔH). This is consistent with the definition of $\Delta G = \Delta H - T\Delta S$. The sign that accompanies ΔG is used to determine the spontaneity of the reaction: a negative sign indicates a spontaneous reaction as written, a positive value indicates a nonspontaneous reaction, and a value of zero indicates a reaction at equilibrium.

Relation of ΔG and the equilibrium constant By mathematical derivation, it can be shown that $\Delta G^\circ = -RT \ln K_e$ or $\Delta G^\circ = -2.303\, RT \log K_e$.

PROBLEMS

1. Calculate the internal energy change, ΔE, for a system when
   (a) $q = -300$ cal; $w = -750$ cal.
   (c) One kcal of heat energy is absorbed by the system and the system does 540 cal of work on the surroundings.

   Soln: (a) The internal energy change of a system ΔE is a balance between heat and work $\Delta E = q - w$. In this case,

   $$\Delta E = -300 - (-750) = 450 \text{ cal}$$

   (c) Heat absorbed by the system is considered positive. Work done by the system is considered positive.

   $$\Delta E = 1000 \text{ cal} - (+540) = +460 \text{ cal}$$

2. For the conditions one liter-atm = 24.2173 cal and the gas constant $R = 1.987$ cal/mole $^\circ$K, what is the value of R in units of liter-atm/mole $^\circ$K?

   Soln: $R = \dfrac{1.987 \text{ cal/mole } ^\circ\text{K} \times 1 \text{ liter-atm}}{24.2173 \text{ cal}}$

   $= 0.08205$ liter-atm/mole $^\circ$K

3. Evaluate in kilocalories the work done when 3.00 moles of helium expand from a volume of 1.00 liter to a volume of 45.0 liters against a pressure of 2.50 atm, generally measured in kcal.

   Soln: Work is defined in pressure-volume units as pressure times the change in volume for one mole.

   $$w = nP(V_2 - V_1) = 3(2.50 \text{ atm}) (45.0 - 1.00) \text{ liters} = 330 \text{ liter-atm}$$

   $$= 330 \text{ liter-atm} \times 24.2 \text{ cal/liter-atm}$$

   $$= 7.99 \times 10^3 \text{ calories, or } 7.99 \text{ kcal}$$

4. How many kcal of heat energy will be liberated when 49.7 grams of manganese are burned to form $Mn_3O_4(s)$ at standard state conditions? ΔH°_{f298} of Mn_3O_4 is equal to -331.7 kcal/mole.

Soln: First write the overall reaction:

$$3Mn(s) + 2O_2(g) \rightarrow Mn_3O_4(s)$$

Next calculate the amount of $Mn_3O_4(s)$ that will form in the reaction.

$$Wt\ Mn_3O_4(s) = \frac{FW\ Mn_3O_4(s) \times 49.7\ g}{3\ (at\ wt\ Mn)}$$

$$= \frac{228.7 \times 49.7\ g}{3(54.9)}$$

$$= 69.0\ g$$

We now find how many moles of Mn_3O_4 are present and multiply by the heat per mole.

$$\Delta H^\circ_{f298\ (Mn_3O_4)} = -331.7\ kcal/mole$$

$$Heat = 69.0\ g \times \frac{1}{228.7\ mole/g} \times -331.7\ kcal/mole$$

$$= -100\ kcal$$

The heat liberated is 100 kcal.

5. (a) Using the enthalpy of formation data in Appendix J, calculate the enthalpy change for the following reactions:
(1) $CaO(s) + SO_3(g) + 2H_2O(l) \rightarrow CaSO_4 \cdot 2H_2O(s)$.
(3) $CaSO_3 \cdot 2H_2O(s) + CO_2(g) \rightarrow CaCO_3(s) + SO_2(g) + 2H_2O(l)$.

Soln: (a) of (1): $\Delta H = \Delta H$ (products) $- \Delta H$ (reactants)

$$\Delta H = \Delta H_{CaSO_4 \cdot 2H_2O(s)} - \Delta H_{CaO(s)} - \Delta H_{SO_3(g)} - 2\Delta H_{H_2O(l)}$$

$$= -483.06 - (-151.9) - (-94.58) - 2(-68.32)$$

$$= -99.9\ kcal$$

(a) of (3): $\Delta H = \Delta H$ (products) $- \Delta H$ (reactants)

$$\Delta H = \Delta H_{CaCO_3(s)} + \Delta H_{SO_2(g)} + 2\Delta H_{H_2O(l)} - \Delta H_{CaSO_3 \cdot 2H_2O(s)} - \Delta H_{CO_2(g)}$$

$$= -288.45 - 70.94 + 2(-68.132) - (-421.2) - (-94.05)$$

$$= 19.2\ kcal$$

6. (a) Calculate the heat (enthalpy) of formation for calcium sulfide, using the enthalpy of formation data in Appendix J, for each of the following reactions:
(a) $CaS(s) + 2O_2(g) + 2H_2O(l) \rightarrow CaSO_4 \cdot 2H_2O(s) + 231.3\ kcal$.

111

Soln: (a) ΔH (reaction) = ΔH (products) − ΔH (reactants)

$-231.3 \text{ kcal} = \Delta H_{CaSO_4 \cdot 2H_2O(s)} - \Delta H_{CaS(s)} - 2\Delta H_{O_2(g)} - 2\Delta H_{H_2O(l)}$

$-231.3 = -483.06 - \Delta H_{CaS(s)} - 2(0) - 2(-68.32)$

$\Delta H_{CaS(s)} = -115.1 \text{ kcal}$

8. Calculate, using the data in Appendix J, the bond energies of N_2, O_2, and NO. All are gases in their most stable form at standard state conditions.

Soln:

Gas	$\Delta H°_{f298}$
N (g)	112.979 kcal/mole
O (g)	59.553 kcal/mole
NO(g)	21.57 kcal/mole

For N(g) and O(g), the above values correspond to one-half the energy required to break the bond in the diatomic molecules N_2 and O_2. Since both diatomic species are assigned values of 0 for ΔH_f, the energy required to form two individual atoms is the same as the bond energy. Thus for the N−N bond:

$2(112.979) = 225.958 = 226.0 \text{ kcal/mole of bonds}$

For the O−O bond:

$2(59.553) = 119.106 = 119.1 \text{ kcal/mole of bonds}$

For NO, the ΔH_f is 21.57 kcal, meaning that heat must be supplied to form the bond. Therefore, the bond in NO is less stable by 21.57 kcal than the sum of the energies required to form N and O atoms. Therefore for NO,

$$\Delta H_f(N) + \Delta H_f(O) - \Delta H_f(NO) = 112.98 + 59.55 - 21.57$$

$$= 150.96$$

$$= 151.0 \text{ kcal/mole of bonds}$$

9. For each of the reactions below, calculate the Gibbs free energy change, the enthalpy change, and the entropy change. Which of the reactions are spontaneous? For which are the entropy changes favorable for the reaction to proceed?
(a) $Fe_2O_3(s) + 13CO(g) \rightarrow 2Fe(CO)_5(g) + 3CO_2(g)$.

Soln: (a) $\Delta G = 2\Delta G_{Fe(CO)_5(g)} + 3\Delta G_{CO_2(g)} - \Delta G_{Fe_2O_3(s)} - 13\Delta G_{CO(g)}$

$= 2(-166.65) + 3(-94.254) - (-177.4) - 13(-32.78)$

$= -333.30 - 282.762 + 177.4 + 426.14$

$= -12.5 \text{ kcal}$

$$\Delta H = 2\Delta H_{Fe(CO)_5(g)} + 3\Delta H_{CO_2(g)} - \Delta H_{Fe_2O_3(s)} - 13\Delta H_{CO(g)}$$
$$= 2(-175.4) + 3(-94.051) - (197.0) - 13(-26.416)$$
$$= -350.8 - 282.153 + 197.0 + 343.408$$
$$= -92.5 \text{ kcal}$$

$$\Delta S = 2\Delta S_{Fe(CO)_5(g)} + 3\Delta S_{CO_2(g)} - \Delta S_{Fe_2O_3(s)} - 13\Delta S_{CO(g)}$$
$$= 2(106.4) + 3(51.06) - 20.89 - 13(47.219)$$
$$= 212.8 + 153.18 - 20.89 - 613.847$$
$$= -268.8 \text{ cal/}^\circ K$$

11. For a certain process at $300^\circ K$, $\Delta G = -18.4$ kcal and $\Delta H = -13.6$ kcal. Find the entropy change for this process at this temperature.

   Soln: The entropy is related to the free energy and enthalpy by the equation:

$$\Delta G = \Delta H - T\Delta S$$
$$-18.4 \text{ kcal} = -13.6 \text{ kcal} - 300^\circ \Delta S$$
$$-4.8 \text{ kcal} = -300^\circ \Delta S$$
$$\Delta S = \frac{4.8 \text{ kcal}}{300^\circ} = \frac{4800 \text{ cal}}{300^\circ}$$
$$= 16.0 \text{ cal/}^\circ K$$

12. (a) For the vaporization of bromine liquid to bromine gas, calculate the change in enthalpy and the change in entropy at standard state conditions.
   (b) From the calculations in (a) discuss relative disorder in bromine liquid compared to bromine gas. On the basis of the enthalpy change, state what you can about the spontaneity of the vaporization.
   (c) Calculate the value of ΔG°_{298} for the vaporization of bromine from the data in Appendix J.
   (d) State what you can about the spontaneity of the process from the value you obtained for ΔG°_{298} in (c).
   (e) Calculate the temperature at which liquid and gaseous Br_2 are in equilibrium with each other at 1 atm (assume ΔH° and ΔS° are independent of temperature).
   (f) From this temperature value (part e) state in which direction the process would be spontaneous.
   (g) Compare ΔH°, ΔS°, and ΔG° in terms of their usefulness in predicting spontaneity of the vaporization of Br_2.

Soln: (a) $Br_2(l) \rightarrow Br_2(g)$

$$\Delta H^\circ_{298} = \Delta H^\circ_{f\, Br_2(g)} - \Delta H^\circ_{f\, Br_2(l)}$$
$$= 7.387 - 0 = 7.387 \text{ kcal/mole}$$

$\Delta S^\circ_{Br_2(l),\, 298}$ = absolute entropy for $Br_2(l)$ at 298°

$$= S^\circ_{Br_2(l),\, 298} - S^\circ_{Br_2(s),\, 0}$$
$$= 36.384 - 0$$
$$= 36.384 \text{ cal/mole } ^\circ K$$

$\Delta S^\circ_{Br_2(g),\, 298}$ = absolute entropy for $Br_2(g)$ at 298°

$$= S^\circ_{Br_2(g),\, 298} - S^\circ_{Br_2(s),\, 0}$$
$$= 58.641 - 0$$
$$= 58.641 \text{ cal/mole } ^\circ K$$

ΔS° = final state − initial state

$$= \Delta S^\circ_{Br_2(g)} - \Delta S^\circ_{Br_2(l)}$$
$$= 58.641 - 36.384$$
$$= 22.257 \text{ cal/mole } ^\circ K$$

This could also be obtained more simply by subtracting the two entropy changes at 298 directly.

(b) The positive value of S_{298} indicates that the disorder in the gaseous state of Br_2 is greater than the disorder in the liquid state. However, the positive value of the enthalpy indicates that for vaporization to occur, heat must be added to the system. The spontaneity of the vaporization can be determined only through the free energy change for the temperature of interest. See part c.

(c) $\Delta G^\circ_{298} = \Delta G^\circ_{Br_2(g),\, 298} - \Delta G^\circ_{Br_2(l),\, 298}$

$$= 0.751 - 0 = 0.751 \text{ kcal/mole}$$

(d) The process as written is not spontaneous, owing to the positive value of ΔG°.

(e) When equilibrium is established, $\Delta G = 0$. Therefore,

$$\Delta G = 0 = \Delta H - T\Delta S \quad \text{or} \quad T\Delta S = \Delta H.$$

From parts a and c,

$$T = \frac{\Delta H}{\Delta S} = \frac{7387}{22.257} = 331.9^\circ K, \text{ or } 58.7^\circ C$$

(f) The process will be spontaneous as T increases.

(g) Only the free energy function can determine the sponeaneity of the reaction. If, however, ΔS is positive and ΔH is negative, the reaction will occur spontaneously.

16. (a) The value of K_e is 9.23×10^{-22} at 25°C for the reaction $CO_2(g) + C(s) \rightarrow 2CO(g)$. What is $\Delta G°_{298}$ for this reaction, calculated from K_e?
(b) Calculate $\Delta G°_{298}$ again, this time using the data in Appendix J. Compare with your answer from (a).

Soln: (a) Using the equation $\Delta G = -2.303 \, RT \log K_e$, we have

$$\Delta G° = -2.303(1.987 \text{ cal/mole °K}) (298°\text{K}) \log 9.23 \times 10^{-22}$$

$$= -2.303(1.987)(298)(-21.03)$$

$$= 28.7 \text{ kcal}$$

(b) $\Delta G° = 2\Delta G°_{CO(g)} - \Delta G°_{CO_2(g)} - \Delta G°_{C(s)}$

$$= 2(-32.780) - (-94.254) - 0$$

$$= 28.7 \text{ kcal}$$

18. (a) If you wished to decompose $CaCO_3(s)$ into $CaO(s)$ and $CO_2(g)$ at atmospheric pressure, what would be the minimum temperature at which you would conduct the reaction?
(b) Is your answer in (a) compatible with Section 17.14 of the text?
(c) Calculate the equilibrium vapor pressure of $CO_2(g)$ above $CaCO_3(s)$ in a closed container at 298°K and 1.00 atm.

Soln: (a) The minimum temperature at which decomposition can occur is the equilibrium temperature when $\Delta G = 0$; thus $\Delta H = T \Delta S$. If we assume that ΔH and ΔS are independent of temperature, $\Delta H° = T \Delta S°$. The reaction proceeds as $CaCO_3(s) \rightleftharpoons CaO(s) + CO_2(g)$.

$\Delta H° = \Delta H°_{CaO(s)} + \Delta H°_{CO_2(g)} - \Delta H°_{CaCO_3(s)}$

$$= -151.9 - 94.051 - (-288.45)$$

$$= 42.5 \text{ kcal}$$

$\Delta S° = \Delta S°_{CaO(s)} + \Delta S°_{CO_2(g)} - \Delta S°_{CaCO_3(s)}$

$$= 9.5 + 51.06 - 22.2$$

$$= 38.4$$

$$T = \frac{\Delta H°}{\Delta S°} = \frac{42,500}{38.4} = 1107°\text{K}$$

$$= 835°\text{C}$$

(b) Yes

(c) Use is made of the free energy to determine the equilibrium pressure.

$$\Delta G° = \Delta H° - T\,\Delta S° = 42.5 - 298(0.0384) = 31.1 \text{ kcal}$$

$$31,100 \text{ cal} = -2.303(1.987 \text{ cal/°K mole}) (298°\text{K}) \log K_e$$

$$\log K_e = \frac{-31,100}{2.303(1.987)\,(298)} = -22.8$$

$$K_e = \text{antilog} (-22.8) = 1.58 \times 10^{-23} \text{ atm}$$

$$K = p_{CO_2} = 1.58 \times 10^{-23} \text{ atm}$$

$$= 1.58 \times 10^{-23} \text{ atm} \times 760 \text{ mm/atm}$$

$$= 1.20 \times 10^{-20} \text{ mmHg}$$

Since

$$\frac{[CaO]\,[CO_2]}{[CaCO_3]} = K_e$$

$$[CO_2] = \frac{[CaCO_3]}{[CaO]} \times K_e = K$$

Then, $K = p_{CO_2}$ in terms of partial pressures.

22. (a) What is the equilibrium vapor pressure, in mmHg, of $H_2O(g)$ above pure $H_2O(l)$ at 298°K and 1 atm (use Appendix J)? Compare your answer to the value given in Section 10.9, Table 10-1.
(b) At what temperature, °C, would water boil if it were under an external pressure of 23.7 mmHg?

Soln: (a) $\Delta G° = \Delta G°_{H_2O(g)} - \Delta G°_{H_2O(l)}$

$$= -54.634 - (-56.687)$$

$$= 2.053 \text{ kcal}$$

$$2,053 \text{ cal} = -2.303(1.987 \text{ cal/°K mole}) (298 °K) \log K_e$$

$$\log K_e = -1.506$$

$$K_e = 0.031 \text{ atm} = p_{H_2O}$$

$$p_{H_2O} = 0.031 \text{ atm} \times 760 \text{ mm/atm} = 23.7 \text{ mmHg}$$

(b) Since the basis of the above calculation is the value of ΔG at 25°C, the temperature of boiling would be 25°C.

UNIT XIII

Electrochemistry
22

INTRODUCTION

The material of this chapter covers three basic areas: electrolytic cells, voltaic cells, and the way in which emf measurements are used to determine thermodynamic functions. The Nernst equation is introduced to allow for concentration and temperature changes. Two sections at the end of this chapter cover the high-interest areas of fuel cells and solar energy.

FORMULAS AND DEFINITIONS

Anode   The electrode towards which negatively charged ions are attracted; electrons are withdrawn from the electrolytic liquid causing oxidation.

Anions   Negatively charged ions.

Cathode   The electrode towards which positively charged ions are attracted; electrons enter the electrolytic liquid causing reduction.

Cations   Positively charged ions.

Electrolytic cell   A chemical reaction system in which electrical energy is consumed to bring about desired chemical changes. Such chemical changes are by definition nonspontaneous. In the process electrons are forced from an outside source onto the cathode, making it negatively charged; and electrons are withdrawn from the anode making it positively charged.

Electrolysis An oxidation-reduction reaction taking place in an electrolytic cell.

Electromotive series An ordering of the elements by their ability to lose electrons. In reduction potentials, potassium has the largest negative value, -2.925 volts for the reaction $K^+ + e^- \rightleftharpoons K$.

emf An acronym for electromotive force, a force or potential causing an electron flow. It is normally measured in volts.

Faraday's law During electrolysis, 96,487 coulombs (1 faraday) of electricity reduce one gram-equivalent of the oxidizing agent and oxidize one gram-equivalent of the reducing agent. In other words, the amount of substance reacted at each electrode during electrolysis is directly proportional to the quantity of electricity passing through the electrolytic cell.

Nernst equation The equation is defined for reactions and for half-reactions having the general form

$$a A + b B \rightleftharpoons c C + d D$$

and
$$x M + n e^- \rightleftharpoons y N$$

as
$$E = E^0 - \frac{0.059152}{n} \log Q$$

where Q, respectively, is $= \dfrac{[C]^c [D]^d}{[A]^a [B]^b}$ and $Q = \dfrac{[N]^y}{[M]^x}$

E = emf for the reaction or half-reaction;

E^0 = standard electrode potential for the cell reaction or the half-reaction;

n = number of electrons required in the redox transfer process according to the balanced equation or half-reactions.

Standard hydrogen electrode Prepared by bubbling hydrogen gas at 25°C and a pressure of one atmosphere around a platinized platinum electrode immersed in a solution one molar in hydrogen ions.

Standard potential Potential of the electrode measured at 25°C, when the concentration of the ions in the solution is 1 M and the pressure of any gas involved is 1 atm.

Thermodynamic functions and their relation to E° Several relations are possible:
$$\Delta G^\circ = -nFE^\circ$$

$$E^\circ = \frac{RT}{nF} \ln K_e$$

$$= \frac{0.0591}{n} \log K_e$$

where n is the number of electrons transferred and F is the faraday, a constant in units of calories per volt ($F = 23.060$ cal/volt), and the temperature is defined at $298°$K.

Voltaic cells Commonly thought of as batteries. They have as their negative terminal the anode, where oxidation occurs. Reduction still occurs at the cathode, but the cathode is the positive terminal.

PROBLEMS

1. A 1.00-liter volume of a concentrated sodium chloride solution is electrolyzed, chlorine being produced at the anode. Calculate the hydroxide ion concentration in the solution after the solution has been electrolyzed for 30 minutes at 1.0 ampere, assuming that the cell is designed so that no chlorine reacts with sodium hydroxide.

   Soln: The electrolysis of a concentrated NaCl solution is represented by the following net reaction, taken from the textbook (Figure 22-2).

$$2H_2O + 2e^- \rightarrow H_2 + 2OH^-$$
$$2Cl^- \rightarrow Cl_2 + 2e^-$$
$$\overline{2H_2O + 2Cl^- \rightarrow H_2 + Cl_2 + 2OH^-}$$

   The production of one mole of hydroxide ions (OH^-) requires one faraday of charge (one mole of electrons).

$$\text{No. of faradays} = I \times t \times \frac{1 \text{ faraday}}{96,500 \text{ coulombs}}$$

$$= \frac{(1.0 \text{ amp}) (30 \text{ min} \times 60 \text{ sec/min})}{96,500 \text{ coulombs}}$$

$$= 1.9 \times 10^{-2}$$

   Therefore, 1.9×10^{-2} mole of OH^- is produced.

3. How many ampere-hours of electricity are required in the electrolytic refining of 3.00 kg of copper?

   Soln: The reduction of one mole of Cu^{2+} ions to metallic copper requires two moles (2 faradays) of charge. One ampere $\cdot$ hour of charge is a current of 1 amp flowing for 1 hr.

   1.0 amp $\cdot$ hr = (1 amp) (60 min/hr) (60 sec/min) = 3600 coul

$$\text{No. moles Cu} = 3.00 \text{ kg} \times 1000 \text{ g/kg} \times \frac{1 \text{ mole}}{63.5 \text{ g}}$$

$$= 47.24 \text{ moles}$$

No. faradays required = 2(47.24 moles) = 94.48

$$\text{No. amp} \cdot \text{hr} = 94.48 \text{ faradays} \times \frac{96,500 \text{ coul}}{\text{faraday}} \times \frac{1 \text{ amp} \cdot \text{hr}}{3600 \text{ coulombs}}$$

$$= 2530$$

7. How many grams of platinum would be deposited from a solution of Na_2PtCl_4 by a current of 5.00 milliamperes flowing for 7.00 hours?

   Soln: In Na_2PtCl_4, Pt is in the 2+ oxidation state. Therefore, the reduction, $Pt^{2+} + 2e^- \rightarrow Pt$, requires 2 faradays of charge per mole of platinum reduced.

$$\text{No. of faradays} = \frac{(0.005 \text{ amp}) (7.00 \text{ hr}) (3600 \text{ sec/hr})}{96,500 \text{ coulombs}} = 0.0013$$

$$\text{Wt of Pt} = (0.0013 \text{ faraday}) \frac{1 \text{ mole}}{2 \text{ faradays}} \times \frac{195 \text{ g}}{\text{mole}} = 0.127 \text{ g}$$

11. A total of 69,500 coulombs of electricity was required in the electrolytic reduction of 16.7 g of a metal from a solution of its *tripositive* ions. What is the metal?

   Soln: The metal is to be identified from its atomic weight calculated from the data. The reduction of one mole of the metal requires three moles of electrons.

$$M^{3+} + 3e^- \rightarrow M$$

   Wt of M produced by one faraday:

$$\text{No. of F} = 69,500 \text{ coulombs} \times \frac{1 \text{ F}}{96,500} = 0.720 \text{ F}$$

$$\text{Wt per faraday} = \frac{16.7 \text{ g}}{0.720 \text{ F}} = 23.2 \text{ g}$$

   Since the reduction of one mole of M requires three faradays, the atomic weight of M is 3(23.2 g) = 69.6. This atomic weight is approximately the weight of gallium.

13. Calculate the emf for cells made up of the pairs of standard electrodes listed below. Consult Table 22-1 and 22-2 (*textbook*) for the standard electrode potentials. Add half-reactions to obtain cell reactions in such a way as to give positive emf values for each cell. Identify the anode and cathode in each cell. Indicate the direction of the cell reaction in each case.

   (a) Pb^{2+}, Pb and Mn^{2+}, Mn.

   (c) Pt, Sn^{4+}, Sn^{2+} and Pt, $Cr_2O_7^{2-}$, Cr^{3+}, H^+.

Soln: (a) The half-reactions are:

$$Mn^{2+} + 2e^- \rightarrow Mn; \quad E^\circ = -1.18 \, v$$

$$Pb^{2+} + 2e^- \rightarrow Pb; \quad E^\circ = -0.13 \, v$$

The spontaneous reaction is determined from the half-reactions by considering the E° values for each half-reaction and arranging the half-reactions and E° signs in a way to give a positive value for E°_{cell}.

$$Mn + Pb^{2+} \rightarrow Mn^{2+} + Pb$$

$$E^\circ_{cell} = E^\circ_{cathode} + E^\circ_{anode}$$

$$= -0.13 \, V + 1.18 \, V = 1.05 \, V$$

(c) The half-reactions are:

$$Cr_2O_7^{2-} + 14H^+ + 6e^- \rightarrow 2Cr^{3+} + 7H_2O; \quad E^\circ = 1.33 \, V$$

$$Sn^{4+} + 2e^- \rightarrow Sn^{2+}; \quad E^\circ = 0.15 \, V$$

The spontaneous reaction will be:

$$3Sn^{2+} + Cr_2O_7^{2-} + 14H^+ \rightarrow 3Sn^{4+} + 2Cr^{3+} + 7H_2O$$

$$E^\circ_{cell} = E^\circ_{cathode} + E^\circ_{anode} = 1.33 \, V + (-0.15 \, V) = 1.18 \, V$$

16. A lead storage battery has initially 100 g of lead and 100 g of PbO_2 plus excess H_2SO_4. Theoretically, how long could this cell deliver a current of 1.00 ampere, without recharging, if it were possible to operate it so that the reaction goes to completion?

Soln: A battery can theoretically deliver current until one of the electrodes is depleted. The reaction at both electrodes produces lead in the Pb^{2+} oxidation state. Hence, two faradays are required in the oxidation of lead to lead(II) and in the reduction of PbO_2 to lead (II). Which electrode contains the greatest amount of lead?

$$\text{No. moles of Pb} = 100 \, g \times \frac{1 \text{ mole}}{207.2 \, g} = 0.483$$

$$\text{No. moles of } PbO_2 = 100 \, g \times \frac{1 \text{ mole}}{239.2 \, g} = 0.418$$

The PbO_2 electrode (the cathode) will be depleted first and will determine the operating time. The charge produced is 0.418 mole times 2 F per mole, 0.836 F.

$$\text{Time} = (0.836 \, F) \frac{1.0 \text{ amp} \cdot hr}{3600 \text{ coulomb}} (96,500 \text{ coul/F}) = 22.4 \, hr$$

19. A heavy silver wire acts as the anode and a platinum wire as the cathode in 250 ml of a solution containing an unknown concentration of chloride ion. A constant current of 0.150 amp is allowed to flow until formation of silver chloride ceases. This requires five minutes and 25 seconds. What was the concentration of chloride ion in the solution?

Soln: The silver electrode is oxidized to produce Ag^+ ions, which subsequently associate with Cl^- ions to produce AgCl. Since AgCl is only slight soluble in water, we shall assume that all the AgCl produced will precipitate.

$$\text{No. F} = (0.150 \text{ amp}) (325 \text{ sec}) \frac{1 \text{ F}}{96,500 \text{ coulombs}}$$

$$= \text{No. moles Ag oxidized} = \text{no. moles } Cl^- \text{ in } 250 \text{ ml}$$

$$= 5.05 \times 10^{-4} \text{ mole}$$

$$[Cl^-] = \frac{5.05 \times 10^{-4} \text{ mole}}{0.250 \text{ liter}} = 2.02 \times 10^{-3} \text{ M}$$

21. (a) A solution contains 0.0100 mole of nickel(II) iodide (NiI_2) in one liter. This solution is to be electrolyzed using inert electrodes. Nickel is plated out at the cathode; iodine is produced at the anode. Determine the standard emf for the electrolytic reaction. What is the minimum voltage which would have to be applied to cause this electrolysis to occur?

Soln: The reactions at the anode and cathode are:

$$2I^- \rightarrow I_2 + 2e^- ; \quad E° = -(0.5355 \text{ V})$$

$$Ni^{2+} + 2e^- \rightarrow Ni; \quad E° = -0.250 \text{ V}$$

The proposed reaction is nonspontaneous with a standard potential of -0.79 V. Thus, the standard potential for the electrolytic reaction is 0.786 V. The cell potential at the specified concentrations is calculated by applying the Nernst equation.

$$[Ni^{2+}]_{initial} = 0.010 \, M; \quad [I^-]_{initial} = 0.020 \, M$$

$$Ni^{2+} + 2e^- \rightarrow Ni; \quad E° = -0.250 \text{ V}$$

$$E_{Ni} = E° - \frac{0.0592}{2} \log \frac{[Ni]}{[Ni^{2+}]}$$

$$= -0.250 \text{ V} - \frac{0.0592}{2} \log \frac{1}{0.01}$$

$$= -0.250 \text{ V} - 0.03 \log 10^2$$

$$= -0.250 \text{ V} - 0.06 = -0.31 \text{ V}$$

$$I_2 + 2e^- \rightarrow 2I^-; \qquad E^° = 0.54 \text{ V}$$

$$E_I = E^° - \frac{0.0592}{2} \log \frac{[I^-]^2}{[I_2]}$$

$$= 0.5355 \text{ V} - \frac{0.0592}{2} \frac{\log (0.02)^2}{1}$$

$$= 0.5355 \text{ V} - 0.0296 \log 4 \times 10^{-4}$$

$$= 0.5355 \text{ V} - 0.0296 \, (-3.40) = 0.64 \text{ V}$$

$$E_{cell} = -0.31 \text{ V} + (-0.64 \text{ V}) = -0.945 \text{ V}$$

To reverse the natural course of reaction, 0.945 V is needed.

24. The solubility of lead(II) sulfate is 1.3×10^{-4} M at 25°C. Calculate the electrode potential of a lead wire in a saturated solution of lead(II) sulfate.

Soln: The Pb^{2+} ion concentration is 1.3×10^{-4} M, and the electrode potential of the cell is calculated from the Nernst equation as:

$$Pb^{2+} + 2e^- \rightarrow Pb; \qquad E^° - 0.13 \text{ V}$$

$$E = -0.13 \text{ V} - \frac{0.0592}{2} \log \frac{1}{1.3 \times 10^{-4}}$$

$$= -0.13 \text{ V} - 0.0296 \text{ V} \, (+3.89) = -0.24 \text{ V}$$

26. Calculate the value for the Gibbs free energy change and the equilibrium constant under standard conditions for the following reactions:
(a) $Mn + Zn^{2+} \rightarrow Zn + Mn^{2+}$; (emf = +0.42 volt).

Soln: (a) The cell potential for the reaction as indicated is 0.42 volt. Free energy and equilibrium values can be calculated from the relations:

$$\Delta G^° = -nFE^° \qquad \text{and} \qquad \Delta G^° = -2.3 \, RT \log K$$

$$\Delta G^° = -(2)(23,061 \text{ cal/V})(0.42 \text{ V}) = -19,371 \text{ cal}$$

$$= -19 \text{ kcal}$$

and

$$\log K = \frac{\Delta G^°}{-2.3 \, RT}$$

$$= \frac{-19,371 \text{ cal}}{-2.3(1.99 \text{ cal/mole } °K)298°K} = 14.2$$

$$K = 10^{14.2} = 1.6 \times 10^{14}$$

30. The standard reduction potentials for the reactions $Ag^+ + e^- \rightarrow Ag$ and $AgCl + e^- \rightarrow Ag + Cl^-$ are 0.7991 and 0.222 volt, respectively. From these

data and the Nernst equation, calculate a value for the solubility product constant (K_{sp}) for AgCl. Compare your answer to the value given in Appendix E.

Soln: The solubility product constant for AgCl is given by

$$AgCl \rightleftharpoons Ag^+ + Cl^-; \quad K_{sp} = [Ag^+] [Cl^-]$$

Rearranging the two electrode potentials gives the desired reaction:

$$Ag \rightarrow Ag^+ + e^- \qquad E^\circ = -0.7991 \text{ V}$$

$$AgCl + e^- \rightarrow Ag + Cl^- \qquad E^\circ = 0.222 \text{ V}$$

$$\overline{AgCl \rightleftharpoons Ag^+ + Cl^-}$$

The ionic conentrations of Ag^+ and Cl^- at equilibrium can be related to the electrode potentials through the appropriate Nernst equations. At equilibrium, the opposing reduction potentials are equal, since ΔG° is zero at equilibrium.

$$Ag^+ + e^- \rightarrow Ag; \quad E = 0.7991 - 0.059 \log \frac{1}{[Ag^+]}$$

$$AgCl + e^- \rightarrow Ag + Cl^-; \quad E = 0.222 - 0.059 \log [Cl^-]$$

Therefore,

$$0.799 - 0.059 \log \frac{1}{[Ag^+]} = 0.222 - 0.059 \log [Cl^-]$$

Transposing terms gives

$$0.577 = 0.059 \left(\log \frac{1}{[Ag^+]} - \log [Cl^-] \right)$$

and

$$0.577 = 0.059 \log \frac{1}{[Ag^+][Cl^-]} = 0.059 \log \frac{1}{K_{sp}}$$

$$\frac{0.577}{0.059} = 9.78 = \log \frac{1}{K_{sp}}$$

$$\log K_{sp} = -9.78$$

$$K_{sp} = 10^{-9.78} = 1.66 \times 10^{-10}$$

This is compared with the value given in the book, 1.8×10^{-10}.

UNIT XIV

Nuclear Chemistry

30

INTRODUCTION

Certain atomic nuclei are unstable and spontaneously disintegrate with the release of energy and penetrating radiations to produce new nuclei. This phenomenon was discovered in 1896 and is known as radioactivity. Since the discovery of radioactivity, scientists in many disciplines have been intensely interested in the structure of atoms including atomic nuclei, and in the application of radioactive isotopes to chemical, biological, and other types of research activities.

The purpose of this chapter is to acquaint you with phenomena related to nuclear change. Specifically, characteristics of nuclear change including natural and artificial nuclear reactions, nuclear stability, methods for detecting and accelerating particles, rates of nuclear reactions, and some research and technological applications of radioactive isotopes are treated.

FORMULAS AND DEFINITIONS

Binding energy (B)   The mass of a nucleus is always less than the combined mass of its constituent particles. The mass difference, mass defect, is related to energy through the equation $E = mc^2$. This energy, called the binding energy of the nucleus, is the energy required to break up the nucleus into its constituent particles. The binding energy is calculated from the equation in which the mass difference is in atomic mass units, amu:

$$B = (ZM_H + (A - Z)M_n - M)c^2$$

where 1 amu = 1.66×10^{-24} g;

Z = atomic number of the element;

M_H = mass of a hydrogen atom, 1.0080 amu;

A = mass number of the nuclide;

M_n = mass of a neutron, 1.0087 amu;

M = measured mass of the nuclide;

c = velocity of light, 3×10^{10} cm/sec.

Energy equivalences   Energy produced by conversion of a mass equivalent to 1 amu, or 1.66×10^{-24} g, to energy through the Einstein equation $E = mc^2$ is

$$E = (1.66 \times 10^{-24} \text{ g}) (3 \times 10^{10} \text{ cm/sec})^2 = 1.49 \times 10^{-3} \text{ erg}$$

$$1 \text{ erg} = 6.24 \times 10^{11} \text{ electron volts}$$

$$= 6.24 \times 10^5 \text{ MeV}$$

$$E = (1.49 \times 10^{-3} \text{ erg})(6.24 \times 10^5 \text{ MeV/erg}) = 931 \text{ MeV}$$

Half-life ($t_{1/2}$)   Amount of time required for one-half the number of nuclei in a radioactive sample to decay to new nuclei. Half-life values range from microseconds to billions of years.

Isotopic mass   Experimentally measured mass of an isotope (nuclide) relative to carbon-12.

Mass number (A)   Sum of the number of neutrons and protons in the nucleus of a specific nuclide.

Nuclear reaction rate   The rate of a nuclear reaction is kinetically of the first order. That is, the number of nuclei of a specific nuclide remaining in a sample after an elapsed time t is a function of the initial number of nuclei N_0 and the half-life of the nuclide. The value N_t, or the number of nuclei at time t, is related to N_0 by the equation $N_t = N_0 \, e^{-kt}$, where k is a proportionality constant called the decay constant. This equation is conveniently rearranged as:

$$\log \frac{N_0}{N_t} = \frac{kt}{2.303}$$

Packing fraction   Quantity related to the binding energy of a nucleus:

$$\text{Packing fraction} = \frac{\text{isotopic mass} - \text{mass number}}{\text{mass number}}$$

PROBLEMS

1. The isotopic mass of $_{13}^{27}$Al is 26.98154. (a) Calculate its binding energy per nucleon (nuclear particle) and compare the value with that given in Fig. 30-2.

Soln: (a) The binding energy and the packing fraction for this nuclide are calculated from the appropriate equations as follows:

$$B = (\text{calculated mass} - \text{measured mass})c^2$$

$$= (ZM_H + (A - Z)M_n - M)c^2$$

$$= (13(1.0080) + (27 - 13)(1.0087) - 26.9815)c^2$$

$$= (13.1040 + 14.1218 - 26.9815)c^2$$

$$= 0.2443 \text{ amu } (c^2)$$

Convert 0.2443 amu to grams, multiply by $(3 \times 10^{10} \text{ cm/sec})^2$ to obtain energy in ergs, and then convert ergs to MeV.

$$\text{Mass} = 0.2443 \text{ amu} \times 1.66 \times 10^{-24} \text{ g/amu} = 4.06 \times 10^{-25} \text{ g}$$

$$B = (4.06 \times 10^{-25} \text{ g})(3 \times 10^{10} \text{ cm/sec})^2 = 3.64 \times 10^{-4} \text{ erg}$$

$$B \text{ (MeV)} = (3.64 \times 10^{-4} \text{ erg})(6.24 \times 10^5 \text{ MeV/erg}) = 227 \text{ MeV}$$

The value of B can be obtained directly from the mass difference by taking advantage of the energy identity, 1 amu = 931 MeV:

$$B = (0.244 \text{ amu})(931 \text{ MeV/amu}) = 227 \text{ MeV}$$

$$\text{Packing fraction} = \frac{26.98154 - 27}{27} = -6.8 \times 10^{-4}$$

(b) $\dfrac{B}{\text{nucleon}} = \dfrac{227 \text{ MeV}}{27 \text{ nucleons}} = 8.41 \text{ MeV}$

2. What percentage of $_{82}^{212}$Pb remains of a 1.00-gram sample, 1.0 minute after it is formed (half-life of 10.6 seconds)? 10 minutes after it is formed?

Soln: The radioactivity of a sample is a function of the number of nuclei in the sample and time. The number of nuclei in a sample determines its mass, its concentration, and its activity. Therefore, the formula

$$\log \frac{N_0}{N_t} = \frac{kt}{2.303}$$

is equally valid for concentration, activity, and mass.

$$\log \frac{C_0}{C_t} = \log \frac{A_0}{A_t} = \log \frac{M_0}{M_t} = \frac{kt}{2.303}$$

The time reference for t and k in the equation must be consistent. In this case, the half-life is given in seconds and the experimental time period in minutes. The same unit must be used for both measurements to satisfy the above equations:

$$k = \frac{0.693}{10.6 \text{ sec}} = \frac{0.0654}{\text{sec}} = 0.0654 \text{ sec}^{-1}$$

$$\log \frac{M_0}{M_t} = \frac{0.0654 \text{ sec}^{-1} \, t}{2.303}$$

$$= \frac{0.0654 \text{ sec}^{-1} \, (1 \text{ min} \times 60 \text{ sec/min})}{2.303}$$

$$= 1.704$$

$$\log M_0 - \log M_t = 1.704$$

Since $M_0 = 1.0$ g, the $\log M_0 = 0$ and

$$\log M_t = -1.704$$

$$M_t = 10^{-1.704} = 0.02 \text{ g at } 1.0 \text{ min}$$

$$\% = \frac{0.020 \text{ g}}{1.0 \text{ g}} \times 100 = 2.0\%$$

3. The isotope of Tl with mass number 208 undergoes beta decay with a half-life of 3.1 min.
   (a) What isotope is the product of the decay?
   (b) Is ^{208} Tl more stable or less stable than an isotope with a 54.5-sec half-life?
   (c) How long will it take for 99.0% of a sample of pure ^{208} Tl to decay?

Soln: (a) $^{208}_{81}\text{Tl} - ^{0}_{-1}B = ^{208}_{82}\text{Pb}$

(b) Tl-208 with a half-life of 3.1 min is considered to be more stable than a nuclide with a 54.5-sec half-life.

(c) The percentage of a sample remaining after a period of time is independent of mass. If 99 percent of the sample has decayed, 1 percent, or a fraction of 0.01 of the original remains.

$$k = \frac{0.693}{3.1 \text{ min}} = 0.224 \text{ min}^{-1}$$

$$\log \frac{M_0}{M_t} = \frac{0.224 \text{ min}^{-1}}{2.303} t = 0.097 \text{ min}^{-1} t$$

$$t = 10.3 \text{ min} \log \frac{1}{0.01} = 10.3(-\log 0.01)$$

$$= (10.3 \text{ min})(2) = 20.6 \text{ min}$$

4. Calculate the time required for 99.999 percent of each of the following radioactive isotopes to decay: (a) $^{226}_{88}$ Ra (half-life, 1590 years).

Soln: The calculations required in parts 4a, 4b, and 4c are similar to those in 3c and 3d.

(a) Consider the initial amount of Ra-226 to be 1.00. Therefore, 0.001 percent, or a fraction of 0.00001, remains. Calculate k and t as in 3c.

$$k = \frac{0.693}{1590y} = 4.36 \times 10^{-4} y^{-1}$$

$$\log \frac{1}{0.00001} = \frac{4.36 \times 10^{-4}}{2.303} y^{-1} t$$

$$= 1.89 \times 10^{-4} y^{-1} t$$

$$t = (5.28 \times 10^3 y)(-\log 0.00001)$$

$$= (5.28 \times 10^3 y)(5)$$

$$= 26,400 y$$

5. The isotope $^{90}_{38}$ Sr is an extremely hazardous isotope in the fallout from a nuclear fission explosion. A 0.500-g sample diminishes to 0.393 g in 10 years. Calculate the half-life.

Soln: Sufficient information is given in the problem to calculate the value of k, which in turn can be used to calculate the half-life of Sr-90.

$$\log \frac{M_0}{M_t} = \frac{kt}{2.303}$$

$$\log \frac{0.500 \text{ g}}{0.393 \text{ g}} = \frac{k(10y)}{2.303} = k(4.34y)$$

$$\log 1.27 = 4.34 y(k)$$

$$k = \frac{0.105}{4.34y} = 0.024 y^{-1}$$

$$\text{half-life} = \frac{0.693}{0.024 y^{-1}} = 28.8 y$$

PART TWO

Exponential Notation and Logarithms

Part Two is a programmed unit designed as a comprehensive review of algebraic operations involving exponential numbers and logarithms. After a careful study of this unit, you should be able to:

1. Write numbers in scientific notation or exponential form.
2. Multiply and divide numbers written in scientific notation.
3. Raise a number to a specified power.
4. Find a specified root of a number.
5. Perform logarithmic operations to calculate products, quotients, powers, and roots.

   Make certain you are able to answer, correctly, all parts of the question in a frame before proceeding to the next frame. Answers to questions in Part Two—Exponential Notation and Logarithms begin on page 146.

PART TWO — EXPONENTIAL NOTATION

[1] Scientific study involves writing and using numbers in the base 10 system. The use of exponential notation enables us to write very large and very small numbers with base 10 in a convenient manner. For example, 0.0000000007 can be written 7×10^{-10} and 7,000,000,000,000,000 can be written 7×10^{15}. Operations with exponential numbers are easily performed with practice. Let us begin our study with some moderate-sized numbers as examples. The number 20 can be written 2×10^1; 40 can be written $4 \times$ _____; 400 can be written $4 \times$ _____.

- -

[2] The number 100 can be thought of as 10×10 or as $10^1 \times 10^1$, both of which equal 10^2. So the number 600 can be written 6×10^2. The number 750 can be written $7.5 \times$ _____. The number 625 can be written _____ $\times$ _____.

- -

[3] Just as 100 equals 10^2, 1000 equals $10 \times 10 \times 10$, or $10^1 \times 10^1 \times 10^1$, or 10^3. The number 4000 can be written 4×10^3. Rewrite the following numbers.
(a) 7700 is 7.7×10^2.
(b) 8250 is 8.25×10^2.
(c) 6255 is _____ $\times$ _____.

- -

[4] The number 100 is 10^2, 1000 is 10^3, 10,000 is 10^4, and 1,000,000 is 10^6. These numbers have a common factor. That is, 100 has two zeros and the exponent of 10 is 2; 100,000 has five zeros and the exponent of 10 is 5. It follows that the exponent of 10 for the number 1,000,000,000 should be _____.

- -

[5] Any number or expression with zero as an exponent has a value of one. For example: $10^0 = 1$; $a^0 = 1$; $(ab)^0 = 1$; etc. Calculate the values for the following exercises.
(a) $5^0 =$ ____
(b) $100^0 =$ ____
(c) $5^0 \times 10^0 =$ ____
(d) $a^0 \times 10^0 \times b^0 =$ ____

- -

[6] In general, a number written in scientific, or exponential, form should be written with one nonzero digit to the left of the decimal point and all digits to the right of the decimal point that are accurately known. Experimentally determined digits are referred to as significant figures. For example: 2254. contains 4 digits that have been experimentally determined and can be written exponentially as 2.254×10^3. The number 254,000 would be written exponentially as 2.54×10^5, since the 3 zeros are place holders. Complete the following exercises:
(a) $7655 = 7.655 \times 10^?$.
(b) $2,225,000,000 = 2.225 \times 10^?$.
(c) $786,000 = 7.86 \times 10^?$.

- -

[7] Write the following numbers in exponential form:
(a) 2250 _____
(b) 60,200,000,000 _____
(c) 5,575,000,000 _____

[8]   Basically the same procedure as that outlined in frame 6 can be used to rewrite very small numbers in exponential form. Examine the accompanying group of numbers and note that the exponent of 10 has a negative sign and represents the number of digits from the decimal point to the right of the first nonzero digit.

$$0.1 = 1 \times 10^{-1}$$
$$0.01 = 1 \times 10^{-2}$$
$$0.001 = 1 \times 10^{-3}$$
$$0.0001 = 1 \times 10^{-4}$$

With this concept in mind, study the following examples and rewrite the exercises:

$$0.00000001 = 1 \times 10^{-8}$$
$$0.00025 = 2.5 \times 10^{-4}$$

(a) $0.000007 = 7 \times 10^{?}$.
(b) $0.0000425 = 4.25 \times 10^{?}$.
(c) $0.0000608 = $ _____

- -

[9]   By the same reasoning as in frame 8, the exponent of 10 for 0.000007 is _____
_____ and for 0.0000425 is _____.

- -

[10]   The number 0.00000235 is correctly written 2.35×10^{-6}, and 506,000 is 5.06×10^{5}. That is, the exponential form of each number contains all the significant figures in the number with one nonzero digit written to the left of the decimal point. In general, the exponent of 10 equals the number of digits from the decimal point of the number to the right of the first nonzero digit. The sign of the exponent is "+" if the number is greater than 1 and "−" if the number is less than 1. For the number 0.00000235, there are 6 digits from the decimal point to the right of 2, and for 506,000 there are 5 digits from the right of 5 to the last digit. (No answer required.)

- -

[11]   Write the following numbers in scientific notation.
(a) 0.003 _____
(b) 0.0075 _____
(c) 0.0002354 _____
(d) 0.0000072 _____

- -

[12]   Writing numbers in a different form is usually done for convenience; care must be taken, however, not to change the magnitude of the number. The change in the value of the exponent must reflect the change in position of the decimal point. For base 10 numbers, if the decimal point is moved 3 places to the left, the exponent of 10 must be increased by the same amount; it is decreased when the decimal point is moved to the right. Consider the following examples:

The number 2.25×10^{2} can be written 22.5×10^{1} or 225×10^{0} without changing the value of the original number. Also, 0.00075 can be written 7.5×10^{-4} or 75×10^{-5}. Using this concept, rewrite the following numbers as indicated.
(a) $235,000 = 23.5 \times 10^{?} = 2.35 \times 10^{?}$
(b) $0.00068 = 6.8 \times 10^{?} = 68 \times 10^{?}$

- -

[13] The implication of the concept discussed in frame 12 is that all real base 10 numbers can be written as the product of a rational number and an exponential term. Therefore, it follows that common arithmetical operations, such as multiplication and division, can be performed with numbers written in exponential form. Let us first consider the rules for multiplying exponential numbers. In multiplying expressions having the same number base system (in our case, base 10), the exponent of the product is merely the sum of the exponents of the multipliers. For example,

$$10^2 \cdot 10^3 = 10^{2+3} = 10^5 = \underline{100,000}$$

Determine the following products:
(a) $10^1 \cdot 10^5 = $ _____
(b) $10^3 \cdot 10^9 = $ _____
(c) $10^4 \cdot 10^{10} = $ _____

- -

[14] The same rule applies when the exponents are negative. For example,

$$10^{-2} \cdot 10^{-3} = 10^{-5}$$

Determine the following products:
(a) $10^{-8} \cdot 10^{-10} = $ _____
(b) $10^{-5} \cdot 10^{-3} = $ _____
(c) $10^{-5} \cdot 10^{-5} = $ _____

- -

[15] When both positive and negative exponents are present in the same expression, the same rule applies. Merely add the exponents algebraically. For example,

$$10^{-3} \cdot 10^5 = 10^2$$

(a) $10^{-15} \cdot 10^{12} = $ _____
(b) $10^{-2} \cdot 10^{-3} \cdot 10^6 = $ _____
(c) $10^{14} \cdot 10^{-12} \cdot 10^3 = $ _____

- -

[16] Now consider the possibility of multiplying expressions written in scientific notation form. To find the product, add the exponents of the exponential term and multiply the rational terms. For example,

$$(2 \times 10^5)(4 \times 10^6) = (2 \times 4)(10^5 \times 10^6) = 8 \times 10^{11}$$

Calculate the following:
(a) $(3.2 \times 10^4)(2 \times 10^5) = $ _____
(b) $(4 \times 10^6)(8 \times 10^{-5}) = $ _____
(c) $(8 \times 10^6)(9.1 \times 10^{-2}) = $ _____
(d) $(2 \times 10^3)(3 \times 10^{-5})(4 \times 10^2) = $ _____

- -

[17] The operations involved in dividing exponential expressions are just as simple as multiplication operations. With base 10 numbers (or any numbers having the same base), the exponent of the quotient is the algebraic difference between the exponent of the numerator and the exponent of the denominator. Study the following examples:

$$\frac{a^x}{a^y} = a^{x-y}$$

$$\frac{a^x}{a^{-y}} = a^{x-(-y)} = a^{x+y}$$

Now with numbers:

$$\frac{10^3}{10^2} = 10^{3-2} = 10^1$$

$$\frac{10^5}{10^{-2}} = 10^{5-(-2)} = 10^{5+2} = 10^7$$

(No answer required.)

- -

[18]   Divide as indicated:

(a) $\dfrac{10^4}{10^1} = $ _____

(b) $\dfrac{10^6}{10^2} = $ _____

(c) $\dfrac{10^5}{10^{-5}} = $ _____

(d) $\dfrac{10^{1.2}}{10^{2.4}} = $ _____

- -

[19]   Dividing two expressions written in scientific notation is done in the same way as multiplying. Let us consider dividing 4000 by 200. The numbers can be written 4×10^3 and 2×10^2. Set up the division in the following manner:

$$\frac{4 \times 10^3}{2 \times 10^2}$$

This division involves two factors, the decimal factor (rational term) and the exponential factor. Therefore, two division operations are required to obtain the quotient: first, division of the rational terms, and second, division of the exponential terms.

$$\frac{4}{2} = 2$$

$$\frac{10^3}{10^2} = 10^{3-2} = 10^1$$

The quotient is 2×10^1, or 20. Determine the quotient for each of the following examples.

(a) $\dfrac{5 \times 10^4}{2 \times 10^2} = $ _____

(b) $\dfrac{7.5 \times 10^{-1}}{3.75 \times 10^3} = $ _____

(c) $\dfrac{9.6 \times 10^{13}}{2.4 \times 10^{-12}} = $ _____

- -

135

[20]   Now consider the division of two numbers written in scientific notation in which the rational term in the numerator is smaller than the rational term in the denominator. The procedure for determining the quotient of the following expression is the same as in frame 19: The quotient

$$\frac{2.4 \times 10^1}{4.8 \times 10^2}$$

of 2.4/4.8 is 0.5 and of $10^1/10^2$ is 10^{-1}, which can be combined as 0.5×10^{-1}. This answer is correct but is not conventionally left in this form. The value should be re-written and left as 5×10^{-2}. This type of change was considered briefly in frame 12. Since the expression 5×10^{-2} is the product of two numbers, there is no change in the value of the expression when one factor is multiplied by a number and the other divided by the same number. For 5×10^{-2}, multiply and divide by 100 as follows:

$$5(100) \times \frac{10^2}{100} = 500 \times 10^0 \quad \text{or} \quad 5(10^2) \times \frac{10^2}{10^2} = 500 \times 10^0$$

Using the same logic, compute the quotient for each of the following expressions and leave the quotient in the proper form for scientific notation.

(a) $\dfrac{2.2 \times 10^{-2}}{6.6 \times 10^{-4}} =$ _____

(b) $\dfrac{8.1 \times 10^{14}}{9.0 \times 10^{-15}} =$ _____

(c) $\dfrac{1.42 \times 10^{-3}}{96.2 \times 10^4} =$ _____

- -

Go to SELF-EVALUATION I.

- -

SELF-EVALUATION I

Complete the following self-evaluation of your skills in writing numbers in scientific notation and multiplying and dividing such terms. Answers to Self-evaluation I questions are on page 147. If one of your answers does not agree with the answer listed, refer to the frame whose number appears in brackets with the answer and rework the problem(s) in that frame.

1. Write the number 7,070,000 in scientific notation.

2. Write the number 2.62×10^5 in rational form.

3. Write the number 0.0000785 in scientific notation.

4. Compute the product $(2 \times 10^6)(3.5 \times 10^8)$.

5. Compute the quotient $(1.9 \times 10^4)/(2.6 \times 10^5)$ and write the result in proper scientific notation.

6. Compute the product $(3.6 \times 10^{-1})(4.2 \times 10^{-7})(5.9 \times 10^8)$ and leave the answer in scientific notation.

- -

Go to frame [21]

- -

[21] Analyzing chemical data often requires calculations with powers and roots of numbers. We shall consider operations involving numbers written in scientific notation. In this frame, let us consider the general case for raising an exponential term to a power. The rule is shown in the following examples:

$$(a^x)^y = a^{xy}$$
$$(a^2)^3 = a^{(2)(3)} = a^6$$

For base 10 terms, the rule is illustrated by the following:

$$(10^2)^3 = 10^6$$
$$(10^{-2})^4 = 10^{-8}$$

Using the rule given above, compute the following products:
(a) $(10^2)^5 = $ _____
(b) $(10^4)^{-2} = $ _____
(c) $(10^{1/2})^4 = $ _____
(d) $(10^{2.5})^2 = $ _____

- -

[22] Raising a number written in scientific notation to a power follows from the general rules in frame 21. In general, the product of two numbers raised to a power is merely the product of each term raised to the indicated power. Study the examples below and then make the indicated calculations.

$$(2 \times 10^2)^2 = 2^2 \times (10^2)^2 = 4 \times 10^4$$
$$(3 \times 10^3)^3 = 3^3 \times (10^3)^3 = 27 \times 10^9 = 2.7 \times 10^{10}$$
$$(2.5 \times 10^4)^2 = 2.5^2 \times (10^4)^2 = 6.25 \times 10^8$$
$$(3 \times 10^{-4})^3 = 3^3 \times (10^{-4})^3 = 27 \times 10^{-12} = 2.7 \times 10^{-11}$$

Evaluate the following expressions:
(a) $(4 \times 10^2)^3 = $ _____
(b) $(5 \times 10^5)^2 = $ _____
(c) $(2.5 \times 10^{-3})^3 = $ _____

- -

[23] The rule for extracting the root of a number is essentially the same as for raising the number to a power. The rule in frame 22 is rewritten for roots as follows. The root of a product involving two or more factors is merely the product of the root of each factor taken collectively. First, let us consider the extraction of roots for exponential terms. Recall that the square root of a given number is equivalent to the number raised to the power $\frac{1}{2}$ and the cube root of a given number is equivalent to the number raised to the power $\frac{1}{3}$. Study the following examples.

$$\sqrt{10^2} = (10^2)^{1/2} = 10^{2/2} = 10^1$$
$$\sqrt[3]{10^3} = (10^3)^{1/3} = 10^{3/3} = 10^1$$
$$\sqrt[4]{10^{-24}} = (10^{-24})^{1/4} = 10^{-24/6} = 10^{-4}$$

Use the rule for extracting roots to compute the values for the following expressions.

(a) $\sqrt{10^{-6}}$ = _____

(b) $\sqrt{10^4}$ = _____

(c) $\sqrt[3]{10^{-6}}$ = _____

(d) $\sqrt[5]{10^{-25}}$ = _____

- -

[24] The exercises in frame 23 involved extracting roots of numbers whose exponents were evenly divisible by the root. For cases in which this is not true, the exponential term must be rewritten as a product of two numbers in which the rational term is greater than 1 and the exponent is evenly divisible by the root. As an example, $\sqrt{10^3}$ is calculated in the following manner:

$$\sqrt{10^3} = \sqrt{1 \times 10^3} = \sqrt{10 \times 10^2}$$

Now apply the general rule in frame 23:

$$\sqrt{10 \times 10^2} = \sqrt{10} \times \sqrt{10^2} = (10)^{1/2} \times (10^2)^{1/2}$$
$$(10)^{1/2} = 3.16 \quad \text{and} \quad (10^2)^{1/2} = 10^1$$

Therefore, $\sqrt{10^3} = 3.16 \times 10^1$.

By using the procedure in this example calculate the value for each of the following expressions. An electronic calculator or a slide rule is essential for the computation work.

(a) $\sqrt{10^5}$ = _____

(b) $\sqrt[3]{10^7}$ = _____

(c) $\sqrt{10^7}$ = _____

(d) $\sqrt[3]{10^8}$ = _____

- -

[25] Next, let us expand the concept developed in frames 23 and 24 to include numbers written in scientific notation. The cube root of 44,000 is calculated in the following manner. First, the number is written in exponential form having the exponent of the exponential term evenly divisible by 3. That is,

$$44{,}000 = 44 \times 10^3$$

and

$$(44{,}000)^{1/3} = (44 \times 10^3)^{1/3}$$
$$= (44)^{1/3} \times (10^3)^{1/3}$$
$$= 3.53 \times 10^1$$

Apply the above procedure in calculating the following roots:

(a) $\sqrt{2500}$ = _____

(b) $\sqrt[3]{1440}$ = _____

(c) $\sqrt{26{,}200}$ = _____

(d) $\sqrt[3]{18{,}000}$ = _____

[26] Roots of small numbers are calculated in the same manner. Study the following examples and then calculate the roots for the exercises.

$$(0.000074)^{1/2} = (7.4 \times 10^{-5})^{1/2}$$
$$= (74 \times 10^{-6})^{1/2}$$
$$= 8.6 \times 10^{-3}$$
$$(0.00000081)^{1/3} = (8.1 \times 10^{-7})^{1/3}$$
$$= (810 \times 10^{-9})^{1/3}$$
$$= (810)^{1/3} \times (10^{-9})^{1/3}$$
$$= 9.35 \times 10^{-3}$$

(a) $(0.000045)^{1/2} = $ _____

(b) $(0.0003)^{1/2} = $ _____

(c) $(0.00061)^{1/3} = $ _____

(d) $(9.6 \times 10^{-16})^{1/3} = $ _____

- -

[27] Fractional roots, such as $\frac{2}{3}$, 0.40, and $\frac{1}{5}$, will be considered in the discussion of logarithms.

- -

Go to SELF-EVALUATION II.

- -

SELF-EVALUATION II

Complete the following self-evaluation of your skills in raising numbers to powers and in extracting roots of numbers. Answers to Self-evaluation II questions are on page 148. If one of your answers does not agree with the answer listed, refer to the frame whose number appears in brackets with the answer and rework the problem(s) in that frame.

1. The value of $(10^8)^3$ is _____.

2. The value of $(2.5 \times 10^4)^2$ is _____.

3. The cube root of 25,000 is _____.

4. The square root of 36,800 is _____.

5. Calculate the cube root of 0.000065.

6. Calculate the square root of 0.000049.

- -

Go to Part Two — Logarithms and
proceed with frame [28].

- -

PART TWO — LOGARITHMS

[28]   In Part Two — Logarithms, we are concerned with arithmetical operations that involve using exponential numbers. We limit our study to operations involving base 10 numbers. Answers to questions in Part Two — Logarithms are on page 148.
    First, let us define the term logarithm. The logarithm of a number is the exponent to which 10 must be raised to equal the number. For example, $100 = 10^2$; that is, the logarithm (abbreviated "log") of 100 is 2. A general interpretation of logs for base 10 numbers is

$$10^x = N$$

where x is the log of the number N. This statement is conventionally rewritten in a form that is more useful for computational work, as shown.

$$\log_{10} N = x$$

The new statement is read in the following manner: "x is the number to which 10 must be raised to equal N." Or alternatively, "x is the base 10 logarithm of N." Using this definition, one can find logs of numbers such as 10, 100, 0.1, and 0.01 as shown below:

$$\log 0.01 = \log 10^{-2} = -2$$
$$\log 0.1 \ = \log 10^{-1} = -1$$
$$\log 1 \quad = \log 10^{0} \ = \ 0$$
$$\log 10 \ \ = \log 10^{1} \ = \ 1$$
$$\log 100 = \log 10^{2} \ = \ 2$$

Find the log of each of the following numbers.

(a) 100,000 = _____   (b) 10,000 = _____
(c) 0.00001 = _____   (d) 0.000001 = _____

— —

[29]   As illustrated in frame 28, logs of numbers are whole numbers when the numbers in question are integral powers of 10. But what about all other numbers? For instance, 25 is between 10 and 100. Therefore, the exponent to which 10 must be raised to equal 25 is between 1 and 2. The log of 25 also is a number between 1 and 2. Perhaps it is obvious at this point that the log of 25 is 1 plus a decimal fraction. In fact, logs always consist of two parts. Consider the following example:

$$100 = 10^{2.0000}$$

The number 2 is called the characteristic and the decimal fraction is called the mantissa. Four decimal places are used after the 2 to establish the accuracy of the logarithm. The characteristic is defined as the whole number representing the lower limit of the exponential range within which a number is located. That is, 25 lies between 10^1 and 10^2 and the characteristic of 25 is 1; 3500 lies between 10^3 and 10^4 and the characteristic of 3500 is 3. Determine the characteristic of each of the following numbers.

(a) 175 = _____   (b) 7 = _____
(c) 1250 = _____   (d) 657,000 = _____

— —

[30]   Mantissas for numbers have been computed and compiled into tables such as the table of logs inside the back cover of this book. Turn to the log table and locate the

following numbers: the number 20 in the column headed by N and the number 3010 under the heading 0 to the right of 20. The numbers are to be interpreted in the following manner. The number 20 should not be read "twenty" but rather "two-zero," which could represent the numbers 2, 20, 200, 2000, and so on. To use the table, mentally place a decimal point to the right of the first significant figure of the numbers under the heading N, thus assigning values ranging from 1.0 to 9.9 for numbers in the N column. As a result of this assignment all the numbers in the columns headed by 0 thru 9 become decimals ranging from .0000 to .9996.

Now let us determine the logarithm of 250. First determine the characteristic to be 2. Next locate 25 in the N column, and to the right in the 0 column find 3979. The log of $250 = 2.3979$ to four decimal places. Also, $10^{2.3979} = 250$. In the same manner, determine the logarithm for each of the following numbers.

(a) log 3500 = _____
(b) log 27,000 = _____
(c) log 450,000 = _____

---

[31] The N, or number, column of the log table in this book contains two-digit numbers, and to determine logs for numbers with three or more digits, the columns headed by 0 thru 9 must be used as shown in the following example. Suppose we wish to find the log of 225. Since 225 lies between 100 and 1000, its characteristic is 2. Now locate 22 in the N column; and to the right of 22 in the column headed by 5, find 3522. The log of 225 is 2.3522. The columns 0 to 9 effectively expand the N column to a 3-digit column. Now determine logs as indicated below.

(a) log 374 = _____   (b) log 2410 = _____
(c) log 1270 = _____   (d) log 575,000 = _____

---

[32] At this point you are able to find the log of a number; for logs to be useful you must also be able to find the number that corresponds to a given log. This process is called finding the antilogarithm, and the procedure is just the reverse of finding the log of a number. As an example, find the number whose log is 2.9782. First, locate the mantissa, 9782, in the table. The mantissa is found to the right of 95 and in the 1 column. This number should be read "nine five one." Second, the magnitude of the number is established by recalling that 0.9782 is the log of 9.51 and the characteristic, 2, indicates that the decimal point is to be shifted two places to the right. That is, 2.9782 is the log of 951. Study the following two examples and then complete the exercises.

Examples:

$$\log x = 3.8779$$
$$.8779 \text{ corresponds to } 7.55$$
$$x = 7550$$

$$\log x = 6.7042$$
$$.7042 \text{ corresponds to } 5.06$$
$$x = 5,060,000$$

(a) log x = 4.8722   x = _____
(b) log x = 3.7745   x = _____
(c) log x = 2.4955   x = _____

---

[33] The table of logs in this book can be used to find logs of three-digit numbers directly, and a fourth place can be found by using the proportional parts (see last three columns of log table). The log scale on a slide rule is convenient for finding logs and antilogs to three places; this frequently is adequate for chemical calculations. Also, electronic calculators equipped with log circuits will display logs to eight or nine places. The proportional parts table is used in the following manner. In frame 31, we found the log of 225. Now find the log of 225.4. Since 225.4 lies between 225 and 226, the log of 225.4 must lie between log of 225 and log of 226, that is, between 2.3522 and 2.3541. To find the log 225.4, go to the 4 column in the proportional parts section of the table on the same line as 22. There find the number 8. To the log 225, namely 2.3522, add 8 to the last digit to get 2.3530. As another example, the log of 3765 is found in a similar manner:

$$\log 3760 = 3.5752 \qquad\qquad \log 3765 = \quad ? \qquad\qquad \log 3770 = 3.5763$$

The proportional parts section, column 5, indicates that 6 must be added to get the log of 3765: log 3765 = 3.5758.
     Find the indicated logs. (a) log 69340 = _____   (b) log 7469 = _____

- -

[34] The proportional parts table also is used to establish the fourth digit in the antilog procedure. In fact, the log-finding procedure is just reversed. The antilog of 2.6458 is found as follows:

$$2.6458 = \log x$$
$$2.6454 = \log 442$$
Diff.   .0004

Locate the 4 in the proportional parts table in the same line as 442. Note that the 4 is in the 4 column. Therefore, 2.6458 is the log of 442.4. Using the same procedure, find the antilogs as indicated.

(a) log x = 1.8989   (b) log x = 2.7791   (c) log x = 3.8429

- -

Go to SELF-EVALUATION III.

- -

SELF-EVALUATION III

Complete the following self-evaluation of your skills in finding logs of numbers and antilogs. Answers to Self-evaluation III questions are on page 149. If one of your answers does not agree with the answer listed, refer to the frame whose number appears in brackets with the answer and rework the problem(s) in that frame.

1. The log of 0.00001 is _____.

2. The log of 125 is _____.

3. The antilog of 2.5502 is _____.

4. By using the proportional parts section of the log table, find the antilog of 1.9677.

5. Find the log of 56.78.

- -

Go to frame [35]

- -

[35]   Next, we consider the common arithmetic operations — multiplication, division, powers, and roots — through the use of logarithms. Rules for performing these operations with exponents were treated in Part Two — Exponential Numbers. Since logs are exponents, the rules for operations with exponents apply to log operations, except that one more step is involved to determine answers. First, we shall treat multiplication.

Recall that the product of two exponential numbers to the same base is the base raised to the sum of the exponents. Therefore, the log of a product is merely the sum of the logs of each factor. In general,

$$\log_{10} (x)(y) = \log_{10} x + \log_{10} y$$

With numbers,

$$\log (10^4)(10^5) = \log (10^4) + \log (10^5)$$
$$= 4 + 5 = 9$$

The log of the product equals 9; therefore the product equals 10^9.

For the following indicated products, compute the logs.

(a) log (44) (51) = _____   (b) log (685) (72) = _____
(c) log (65) (7850) = _____   (d) log (45) (421) (7850) = _____

- -

[36]   Products can be computed by calculating the log of a product, then finding the antilog. The table of proportional parts is invaluable at this point. The product of two numbers, such as (75) (85), is computed in the following manner:

$$\log (75)(85) = \log 75 + \log 85$$
$$= 1.8751 + 1.9294$$
$$= 3.8045$$

Therefore, $(75)(85) = 10^{3.8045}$, or from the antilog, 6376. If this product is computed by long-hand multiplication, its value is found to be 6375. The apparent discrepancy is due to rounding in the evaluation of logs. However, answers generally will be accurate to the number of allowable significant figures. Use the above procedure to compute the following products.

(a) (44) (51)   (b) (685) (72)   (c) (65) (7850)   (d) (45) (421) (7850)

- -

[37]   All examples involving logs encountered thus far have dealt with numbers greater than 1. The use of logs is confined to all positive numbers, that is, numbers greater than zero. Now consider finding the log of a number whose value is greather than zero and less than one. As an example, the log of 0.5 is found in the following manner. First, rewrite the number in scientific notation as 5×10^{-1}. Now compute the log of the product.

$$\log 5 \times 10^{-1} = \log 5 + \log 10^{-1}$$
$$= 0.6990 + (-1) = -0.301$$

Therefore $0.5 = 10^{-0.301}$.

Compute the following logs:

(a) log 0.0065   (b) log 0.0043   (c) log 0.000675

- -

[38]   Finding the antilog in which the log is negative requires one more step than if the log is positive. Although several methods can be used for this procedure, we consider one

method, indicated in the following example: Find the number whose log is -2.1871.
Let x be the number in question. Therefore,

$$\log x = -2.1871$$
$$x = 10^{-2.1871}$$

Since all log values in the table are positive, the log must be rewritten as the sum of two numbers of which one is an integral power of 10 and the other is greater than zero but less than 1, their sum equaling the value of the log. For the above case, the log can be written

$$10^{-2.1871} = 10^{3-2.1871} \times 10^{-3}$$
$$= 10^{0.8129} \times 10^{-3}$$

In other words, what value can be added to -3, the next-higher negative number, to get -2.1871? The value of x is the antilog of 0.8129 times 10^{-3}: 6.50×10^{-3}.
Find the antilog of the following logs.

(a) -1.6884 (b) -3.6590 (c) -4.8652

---

[39] Division operations with logs are in a sense opposite to multiplication operations; that is, as the log of the product of two terms is the sum of the logs, so the log of the quotient of two terms is the difference of the logs. In general,

$$\log_{10} \frac{x}{y} = \log_{10} x - \log_{10} y$$

$$\log \frac{10^4}{10^2} = \log 10^4 - \log 10^2 = 4 - 2 = 2$$

$$\log \frac{4}{2} = \log 4 - \log 2$$
$$= 0.6021 - 0.3010 = 0.3011$$

Then $\frac{4}{2} = 10^{0.3011} = 2$.
Applying the above procedure, calculate the quotient for the following exercises.

(a) $\dfrac{4650}{321}$ (b) $\dfrac{69,500}{225}$

---

[40] Now consider a division operation with logs in which the numerator is smaller than the denominator. As an example, the quotient 2/4 is calculated in the following manner.

$$\log \frac{2}{4} = \log 2 - \log 4$$
$$= 0.3010 - 0.6021$$
$$= -0.3011$$

Then

$$\frac{2}{4} = 10^{-0.3011}$$
$$= 10^{0.6989} \times 10^{-1}$$

The antilog of 0.6989 is 5 and the quotient of 2/4 is 5×10^{-1}, or 0.5.
Calculate the following quotients by using logs.

(a) $\dfrac{425}{575}$ (b) $\dfrac{1650}{18,250}$ (c) $\dfrac{(465)\,(830)}{960}$

---

[41]   Last, we consider the use of logs for raising numbers to powers and for extracting roots of numbers. Raising numbers to powers greater than 2 or 3 becomes cumbersome and time-consuming if one does not use a calculator or logs. Logs are exponents and the rules for raising exponential terms to a power or root apply to logs. That is,

$$\log_{10} (N)^x = x \log_{10} N$$

With numbers, we have

$$\log_{10} (2)^5 = 5 \log_{10} 2$$
$$= 5(0.3010) = 1.5050$$

Then

$$(2)^5 = 10^{1.5050} = 32 \qquad \text{(E)}$$

As another example, the value of $(25)^{2.6}$ is _____.

$$\log (25)^{2.6} = 2.6 \log 25$$
$$= 2.6(1.3979) = 3.6345$$
$$(25)^{2.6} = 10^{3.6345} = 4310$$

Using logs, raise the following numbers to the powers indicated.

(a) $(1250)^4$ (b) $(21)^{3.4}$ (c) $(145)^{8.67}$

- -

[42]   Extracting roots with logs involves exactly the same operations as raising numbers to powers. For example, the cube root of 265 is found in the following manner:

$$\log (265)^{1/3} = \frac{1}{3} \log 265$$
$$= \frac{1}{3}(2.4232) = 0.8077$$
$$\text{antilog of } 0.8077 = 6.42$$

Calculate the indicated roots.

(a) $(475)^{1/5}$ (b) $(1430)^{2/5}$

- -

Go to **SELF-EVALUATION IV**.

- -

SELF-EVALUATION IV

Complete the following self-evaluation of your skills in using logs for multiplication, division, raising numbers to powers, and extracting roots of numbers. Answers to Self-evaluation IV questions are on page 150. If one of your answers does not agree with the answer listed, refer to the frame whose number appears in brackets with the answer and rework the problem(s) in that frame.

1. Calculate the product of (250) (325) (460).

2. Calculate the quotient of 4850/2320.

3. Calculate the cube root of 4365.

4. Calculate the quotient of 275/865.

5. Evaluate $(117)^4$.

ANSWERS FOR PROGRAMMED UNITS

1.

$40 = 4 \times 10 = 4 \times 10^1$

$400 = 4 \times 100 = 4 \times 10^2$

_ _ _ _ _ _ _ _ _ _ _ _ _ _ _ _ _ _

2.

$750 = 7.5 \times 100 = 7.5 \times 10^2$

$625 = 6.25 \times 100 = 6.25 \times 10^2$

_ _ _ _ _ _ _ _ _ _ _ _ _ _ _ _ _ _

3.

(a) $7700 = 7.7 \times 1000 = 7.7 \times 10^3$

(b) $8250 = 8.25 \times 1000 = 8.25 \times 10^3$

(c) $6255 = 6.255 \times 1000 = 6.255 \times 10^3$

_ _ _ _ _ _ _ _ _ _ _ _ _ _ _ _ _ _

4.

Since 1,000,000,000 contains 9 zeros, the exponent of 10 is 9. In exponential notation 1,000,000,000 is written 1×10^9.

_ _ _ _ _ _ _ _ _ _ _ _ _ _ _ _ _ _

5.

(a) $5^0 = 1$

(b) $100^0 = 1$

(c) $5^0 \times 10^0 = 1 \times 1 = 1$

(d) $a^0 \times 10^0 \times b^0 = 1 \times 1 \times 1 = 1$

_ _ _ _ _ _ _ _ _ _ _ _ _ _ _ _ _ _

6.

(a) $7655 = 7.655 \times 10^3$

(b) $2,225,000,000 = 2.225 \times 10^9$

(c) $786,000 = 7.86 \times 10^5$

_ _ _ _ _ _ _ _ _ _ _ _ _ _ _ _ _ _

7.

(a) $2250 = 2.25 \times 10^3$

(b) $60,200,000,000 = 6.02 \times 10^{10}$

(c) $5,575,000,000 = 5.575 \times 10^9$

_ _ _ _ _ _ _ _ _ _ _ _ _ _ _ _ _ _

8.

(a) 7×10^{-6}

(b) 4.25×10^{-5}

(c) 6.08×10^{-5}

_ _ _ _ _ _ _ _ _ _ _ _ _ _ _ _ _ _

9.

$0.000007 = 7 \times 10^{-6}$

$0.0000425 = 4.25 \times 10^{-5}$

_ _ _ _ _ _ _ _ _ _ _ _ _ _ _ _ _ _

10.

No answer required.

_ _ _ _ _ _ _ _ _ _ _ _ _ _ _ _ _ _

11.

(a) 3×10^{-3}

(b) 7.5×10^{-3}

(c) 2.354×10^{-4}

(d) 7.2×10^{-6}

_ _ _ _ _ _ _ _ _ _ _ _ _ _ _ _ _ _

12.

(a) $23.5 \times 10^4 = 2.35 \times 10^5$

(b) $6.8 \times 10^{-4} = 68 \times 10^{-5}$

_ _ _ _ _ _ _ _ _ _ _ _ _ _ _ _ _ _

13.

(a) $10^1 \cdot 10^5 = 10^{1+5} = 10^6$

(b) $10^3 \cdot 10^9 = 10^{3+9} = 10^{12}$

(c) $10^4 \cdot 10^{10} = 10^{4+10} = 10^{14}$

_ _ _ _ _ _ _ _ _ _ _ _ _ _ _ _ _ _

14.

(a) $10^{-8} \cdot 10^{-10} = 10^{-8+(-10)} = 10^{-18}$

(b) $10^{-5} \cdot 10^{-3} = 10^{-5+(-3)} = 10^{-8}$

(c) $10^{-5} \cdot 10^{-5} = 10^{-5+(-5)} = 10^{-10}$

_ _ _ _ _ _ _ _ _ _ _ _ _ _ _ _ _ _

15.

(a) $10^{-15} \cdot 10^{12} = 10^{-15+12} = 10^{-3}$

(b) $10^{-2} \cdot 10^{-3} \cdot 10^6 = 10^{-2+(-3)+6} =$
$10^{-5+6} = 10^1$

(c) $10^{14} \cdot 10^{-12} \cdot 10^3 =$
$10^{14+(-12)+3} = 10^5$

16.

(a) $(3.2 \times 10^4)(2 \times 10^5) =$
$(3.2)(2)(10^4)(10^5) = 6.4 \times 10^9$

(b) $(4 \times 10^6)(8 \times 10^{-5}) =$
$(4)(8)(10^6)(10^{-5}) = 32 \times 10^1$, or
3.2×10^2

(c) $(8 \times 10^6)(9.1 \times 10^{-2}) =$
$(8)(9.1)(10^6)(10^{-2}) = 72.8 \times 10^4$,
or 7.28×10^5

(d) 24×10^0, or 2.4×10^1

17.

No answer required.

18.

(a) $\dfrac{10^4}{10^1} = 10^{4-1} = 10^3$

(b) $\dfrac{10^6}{10^2} = 10^{6-2} = 10^4$

(c) $\dfrac{10^5}{10^{-5}} = 10^{5-(-5)} = 10^{5+5} = 10^{10}$

(d) $\dfrac{10^{1.2}}{10^{2.4}} = 10^{1.2-2.4} = 10^{-1.2}$

19.

(a) $\dfrac{5 \times 10^4}{2 \times 10^2} = \dfrac{5}{2} \times \dfrac{10^4}{10^2} = 2.5 \times 10^{4-2} =$
2.5×10^2

(b) $\dfrac{7.5 \times 10^{-1}}{3.75 \times 10^3} = \dfrac{7.5}{3.75} \times \dfrac{10^{-1}}{10^3} =$
$2 \times 10^{-1-(3)} = 2 \times 10^{-4}$

(c) $\dfrac{9.6 \times 10^{13}}{2.4 \times 10^{-12}} = \dfrac{9.6}{2.4} \times \dfrac{10^{13}}{10^{-12}} =$
$4.0 \times 10^{13-(-12)} = 4.0 \times 10^{25}$

20.

(a) $\dfrac{2.2 \times 10^{-2}}{6.6 \times 10^4} = \dfrac{2.2}{6.6} \times \dfrac{10^{-2}}{10^4} = 0.33 \times 10^{-6}$
should be in form 3.3×10^{-7}, de-
rived as follows:

$(0.33)(10^1) \times \dfrac{10^{-6}}{10^1} = \underline{3.3 \times 10^{-7}}$

(b) $\dfrac{8.1 \times 10^{14}}{9.0 \times 10^{-15}} = \dfrac{8.1}{9.0} \times \dfrac{10^{14}}{10^{-15}} =$
$0.9 \times 10^{29} = 9 \times 10^{28}$, derived by:

$(0.9)(10^1) \times \dfrac{10^{29}}{10^1} = \underline{9 \times 10^{28}}$

(c) $\dfrac{1.42 \times 10^{-3}}{96.2 \times 10^4} = \dfrac{1.42}{96.2} \times \dfrac{10^{-3}}{10^4} =$
$0.0148 \times 10^{-7} = 1.48 \times 10^{-9}$
derived from

$(0.0148)(10^2) \times \dfrac{10^{-7}}{10^2} = \underline{1.48 \times 10^{-9}}$

Self-evaluation I Answers

1. 7.07×10^6 [1 thru 7]

2. 262,000   [6 and 7]

3. 7.85×10^{-5} [8, 9 and 10]

4. 7×10^{14} [16]

5. 7.3×10^{-2} [19] and [20]

6. 8.92×10^1 [20]

21.

(a) $(10^2)^5 = 10^{(2)(5)} = 10^{10}$

(b) $(10^4)^{-2} = 10^{-8}$

(c) $(10^{1/2})^4 = 10^{4/2} = 10^2$

(d) $(10^{2.5})^2 = 10^{(2.5)(2)} = 10^5$

22.

(a) $(4 \times 10^2)^3 = (4^3)(10^2)^3 = 64 \times 10^6 =$
6.4×10^7

(b) $(5 \times 10^5)^2 = (5^2)(10^5)^2 = 25 \times 10^{10} =$
2.5×10^{11}

(c) $(2.5 \times 10^{-3})^3 = (2.5)^3 (10^{-3})^3 =$
$15.6 \times 10^{-9} = 1.56 \times 10^{-8}$

23.
(a) $\sqrt{10^{-6}} = (10^{-6})^{1/2} = 10^{-6/2} = 10^{-3}$

(b) $\sqrt{10^4} = 10^{4/2} = 10^2$

(c) $\sqrt[3]{10^{-6}} = (10^{-6})^{1/3} = 10^{-6/3} = 10^{-2}$

(d) $\sqrt[5]{10^{-25}} = 10^{-5}$

- - - - - - - - - - - - - - - - - - -

24.
(a) $\sqrt{10^5} = (10^5)^{1/2} = (1 \times 10^5)^{1/2} =$
$(10 \times 10^4)^{1/2} = 3.16 \times 10^2$

(b) $\sqrt[3]{10^7} = (1 \times 10^7)^{1/3} = (10 \times 10^6)^{1/3} =$
$(10)^{1/3} \times (10^6)^{1/3} = 2.16 \times 10^2$

(c) $\sqrt{10^7} = (1 \times 10^7)^{1/2} = (10 \times 10^6)^{1/2} =$
3.16×10^3

(d) $\sqrt[3]{10^8} = (1 \times 10^8)^{1/3} =$
$(100 \times 10^6)^{1/3} = (100)^{1/3} \times (10^6)^{1/3} =$
4.65×10^2

- - - - - - - - - - - - - - - - - - -

25.
(a) $\sqrt{2500} = (25 \times 10^2)^{1/2} = 5 \times 10^1$

(b) $\sqrt[3]{1440} = (1.44 \times 10^3)^{1/3} =$
$(1.44)^{1/3} \times (10^3)^{1/3} = 1.13 \times 10^1$

(c) $\sqrt{26{,}200} = (2.62 \times 10^4)^{1/2} =$
1.62×10^2

(d) $\sqrt[3]{18{,}000} = (18 \times 10^3)^{1/3} =$
2.62×10^1

- - - - - - - - - - - - - - - - - - -

26.
(a) $(0.000045)^{1/2} = (45 \times 10^{-6})^{1/2} =$
6.7×10^{-3}

(b) $(0.0003)^{1/2} = (3 \times 10^{-4})^{1/2} =$
1.73×10^{-2}

(c) $(0.00061)^{1/3} = (610 \times 10^{-6})^{1/3} =$
8.48×10^{-2}

(d) $(9.6 \times 10^{-16})^{1/3} = (960 \times 10^{-18})^{1/3} =$
9.86×10^{-6}

- - - - - - - - - - - - - - - - - - -

27.
No answer required.

Self-evaluation II Answers

1. $(10^8)^3 = 10^{24}$ [21]

2. $(2.5 \times 10^4)^2 = (2.5)^2 \times (10^4)^2 =$
6.25 $\times 10^8$ [22]

3. $(25{,}000)^{1/3} = 2.92 \times 10^1$ [25]

4. $(36{,}800)^{1/2} = 1.92 \times 10^2$ [25]

5. $(0.000065)^{1/3} = 4.02 \times 10^{-2}$ [26]

6. $(0.000049)^{1/2} = 7 \times 10^{-3}$ [26]

- - - - - - - - - - - - - - - - - - -

28.
(a) $\log 100{,}000\ = \log 10^5\ =\ \ \ \ 5$

(b) $\log 10{,}000\ \ \ = \log 10^4\ =\ \ \ \ 4$

(c) $\log 0.00001\ = \log 10^{-5}\ =\ \ -5$

(d) $\log 0.000001 = \log 10^{-6}\ =\ \ -6$

- - - - - - - - - - - - - - - - - - -

29.
(a) $10^2 - 175 - 10^3$ $\underline{2}$

(b) $10^0 - 7 - 10^1$ $\underline{0}$

(c) $10^3 - 1250 - 10^4$ $\underline{3}$

(d) $10^5 - 657{,}000 - 10^6$ $\underline{5}$

- - - - - - - - - - - - - - - - - - -

30.
(a) $3500 = 3.5 \times 10^3$
The characteristic is 3. The mantissa
is .5441.
$\log 3500 = 3.5441$ or $10^{3.5441} = 3500$

(b) $27{,}000 = 2.7 \times 10^4$
Characteristic = 4; mantissa = .4314.
$\log 27{,}000 = 4.4314$ or $10^{4.4314} =$
27,000

(c) $450{,}000 = 4.5 \times 10^5$
Characteristic = 5; mantissa = .6532.
$\log 450{,}000 = 5.6532$ or $10^{5.6532} =$
450,000

31.

(a) $374 = 3.74 \times 10^2$
Characteristic = 2; mantissa = .5729
log 374 = 2.5729

(b) $2410 = 2.41 \times 10^3$
Characteristic = 3; mantissa = .3820
log 2410 = 3.3820 or $10^{3.3820} = 2410$

(c) $1270 = 1.27 \times 10^3$
Characteristic = 3; mantissa = .1038
log 1270 = 3.1038 or $10^{3.1038} = 1270$

(d) log 575,000 = 5.7597

– – – – – – – – – – – – – – – – –

32.

(a) $\log x = 4.8722$ or $x = 10^{4.8722}$
$10^{.8722} = 7.45$ so $10^{4.8722} = 74500.$
4 places

(b) $\log x = 3.7745$ or $x = 10^{3.7745}$
$10^{.7745} = 5.95$ so $10^{3.7745} = 5950.$
3 places

(c) $\log x = 2.4955$ or $x = 10^{2.4955}$
$10^{.4955} = 3.13$ so $10^{2.4955} = 313.$
2 places

– – – – – – – – – – – – – – – – –

33.

(a) log 69340 = 4.8407 + .0002 = 4.8409

(b) log 7469 = 3.8727 + .0005 = 3.8732

– – – – – – – – – – – – – – – – –

34.

(a) From the tabled values, 8989 corresponds to 7920 + 3 = 7923. The characteristic is 1; therefore $x = 79.23$.

(b) $x = 601 + .3 = 601.3$

(c) $x = 6960 + 5 = 6965$

– – – – – – – – – – – – – – – – –

Self evaluation III Answers

1. −5 [27]

2. 2.0969 [28]

3. 355 [32]

4. 92.84 [34]

5. 1.7542 [33]

35.

(a) log (44) (51) = log 44 + log 51
log 44 = 1.6435
log 51 = 1.7076
log (44) (51) = 3.3511

(b) log (685) (72) = log 685 + log 72
log 685 = 2.8357
log 72 = 1.8573
log (685) (72) = 4.6930

(c) log (65) (7850) = log 65 + log 7850
log 65 = 1.8129
log 7850 = 3.8949
log (65) (7850) = 5.7078

(d) log (45) (421) (7850) = log 45 +
log 421 + log 7850
log 45 = 1.6532
log 421 = 2.6243
log 7850 = 3.8949
log (45) (421) (7850) = 8.1724

– – – – – – – – – – – – – – – – –

36.

(a) log (44) (51) = log 44 + log 51
= 1.6435 + 1.7076
= 3.3511
(44) (51) = $10^{3.3511}$, or 2244

(b) log (685) (72) = log 685 + log 72
= 2.8357 + 1.8573
= 4.6930
(685) (72) = $10^{4.6930}$, or 49,320

(c) log (65) (7850) = log 65 + log 7850
= 1.8129 + 3.8949
= 5.7078
(65) (7850) = 510,200

(d) log (45) (421) (7850) = 8.1724; then
product = 1.487×10^8.

– – – – – – – – – – – – – – – – –

37.

(a) log 0.0065 = log 6.5×10^{-3}
= log 6.5 + log 10^{-3}
= 0.8129 + (−3)
= −2.1871

(b) log 0.0043 = log 4.3×10^{-3}
= log 4.3 + log 10^{-3}
= 0.6335 + (−3)
= −2.3665

(c) $\log 0.000675 = \log 6.75 \times 10^{-4}$
$= \log 6.75 + \log 10^{-4}$
$= 0.8293 + (-4)$
$= -3.1707$

- - - - - - - - - - - - - - - - - - -

38.
(a) $x = 10^{-1.6884}$
$= 10^{0.3410} \times 10^{-2}$
$= 2.049 \times 10^{-2}$

(b) $x = 10^{-3.6590}$
$= 10^{0.3410} \times 10^{-4}$
$= 2.193 \times 10^{-4}$

(c) $x = 10^{-4.8652}$
$= 10^{0.1348} \times 10^{-5}$
$= 1.364 \times 10^{-5}$

- - - - - - - - - - - - - - - - - - -

39.
(a) $\log \dfrac{4650}{321} = \log 4650 - \log 321$
$= 3.6675 - 2.5065$
$= 1.1610$
Then

$\dfrac{4650}{321} = 10^{1.1610} = 14.48$

(b) $\log \dfrac{69,500}{225} = \log 69,500 - \log 225$
$= 4.8420 - 2.3522$
$= 2.4898$
Then

$\dfrac{69,500}{225} = 10^{2.4898} = 308.8$

- - - - - - - - - - - - - - - - - - -

40.
(a) $\log \dfrac{425}{575} = \log 425 - \log 575$
$= 2.6284 - 2.7597$
$= -0.1313$
Then $\dfrac{425}{575} = 10^{-0.1313} = 10^{0.8687} \times 10^{-1}$
antilog of $0.8687 = 7.391$
$\dfrac{425}{575} = 7.39 \times 10^{-1}$

(b) $\log \dfrac{1650}{18,250} = \log 1650 - \log 18,250$
$= 3.2175 - 4.2613$
$= -1.0438$
$\dfrac{1650}{18,250} = 10^{-1.0438}$
$= 10^{0.9562} \times 10^{-2}$
antilog $= 9.04 \times 10^{-2}$

(c) $\log \dfrac{(465)\,(830)}{960}$
$= \log 465 + \log 830 - \log 960$
$= 2.6675 + 2.9191 - 2.9823$
$= 2.6043$
antilog $= 402.1$

- - - - - - - - - - - - - - - - - - -

41.
(a) $\log (1250)^4 = 4 \log 1250 = $
$4(3.0969) = 12.3876$
$(1250)^4 = 2.44 \times 10^{12}$

(b) $\log (21)^{3.4} = 3.4 \log 21 = $
$3.4(1.3222) = 4.4955$
$(21)^{3.4} = 3.13 \times 10^4$

(c) $(145)^{8.67} = 5.48 \times 10^{18}$

- - - - - - - - - - - - - - - - - - -

42.
(a) $\log (475)^{1/5} = \dfrac{1}{5} \log 475 = $
$\dfrac{1}{5}(2.6767) = 0.5353$
antilog of $0.5353 = 3.43$

(b) $\log (1430)^{2/5} = \dfrac{2}{5} \log 1430 = $
$\dfrac{2}{5}(3.1553) = 1.2621$
antilog of $1.2621 = 18.28$

- - - - - - - - - - - - - - - - - - -

Self-evaluation IV Answers

1. 3.74×10^7 [36]

2. 2.09 [37]

3. 16.34 [42]

4. 0.32 [38] and [39]

5. 1.87×10^8 [41]